POLYMER PROCESSING AND CHARACTERIZATION

Advances in Materials Science

Volume 1

POLYMER PROCESSING AND CHARACTERIZATION

Edited By

Sabu Thomas, PhD, Deepalekshmi Ponnamma and Ajesh K. Zachariah

Apple Academic Press

TORONTO NEW JERSEY

Apple Academic Press Inc.
3333 Mistwell Crescent
Oakville, ON L6L 0A2
Canada

Apple Academic Press Inc.
1613 Beaver Dam Road, Suite # 104
Point Pleasant, NJ 08742
USA

Exclusive worldwide distribution by CRC Press, a Taylor & Francis Group

International Standard Book Number: 978-1-926895-15-4 (Hardback)

Printed in the United States of America on acid-free paper

Library of Congress Control Number: 2012935646

Library and Archives Canada Cataloguing in Publication

Polymer processing and characterization/edited by Sabu Thomas, Deepalekshmi Ponnamma and Ajesh K. Zachariah.
(Recent advances in materials science; 1)

Includes bibliographical references and index.
ISBN 978-1-926895-15-4
1. Polymers. 2. Polymeric composites. I. Thomas, Sabu II. Ponnamma, Deepalekshmi, 1984- III. Zachariah, Ajesh K., 1983- IV. Series: Recent advances in materials science (Toronto); 1

TA455.P58P65 2012 620.1›92 C2012-900043-4

Apple Academic Press also publishes its books in a variety of electronic formats. Some content that appears in print may not be available in electronic format. For information about Apple Academic Press products, visit our website at **www.appleacademicpress.com**

Advances in Materials Science

Series Editors-in-Chief

Sabu Thomas, PhD

Dr. Sabu Thomas is the Director of the School of Chemical Sciences, Mahatma Gandhi University, Kottayam, India. He is also a full professor of polymer science and engineering and Director of the Centre for nanoscience and nanotechnology of the same university. He is a fellow of many professional bodies. Professor Thomas has authored or co-authored many papers in international peer-reviewed journals in the area of polymer processing. He has organized several international conferences and has more than 420 publications, 11 books and two patents to his credit. He has been involved in a number of books both as author and editor. He is a reviewer to many international journals and has received many awards for his excellent work in polymer processing. His h Index is 42. Professor Thomas is listed as the 5th position in the list of Most Productive Researchers in India, in 2008.

Mathew Sebastian, MD

Dr. Mathew Sebastian has a degree in surgery (1976) with specialization in Ayurveda. He holds several diplomas in acupuncture, neural therapy (pain therapy), manual therapy and vascular diseases. He was a missionary doctor in Mugana Hospital, Bukoba in Tansania, Africa (1976-1978) and underwent surgical training in different hospitals in Austria, Germany, and India for more than 10 years. Since 2000 he is the doctor in charge of the Ayurveda and Vein Clinic in Klagenfurt, Austria. At present he is a Consultant Surgeon at Privatclinic Maria Hilf, Klagenfurt. He is a member of the scientific advisory committee of the European Academy for Ayurveda, Birstein, Germany, and the TAM advisory committee (Traditional Asian Medicine, Sector Ayurveda) of the Austrian Ministry for Health, Vienna. He conducted an International Ayurveda Congress in Klagenfurt, Austria, in 2010. He has several publications to his name.

Anne George, MD

Anne George, MD, is the Director of the Institute for Holistic Medical Sciences, Kottayam, Kerala, India. She did her MBBS (Bachelor of Medicine, Bachelor of Surgery) at Trivandrum Medical College, University of Kerala, India. She acquired a DGO (Diploma in Obstetrics and Gynaecology) from the University of Vienna, Austria; Diploma Acupuncture from the University of Vienna; and an MD from Kottayam Medical College, Mahatma Gandhi University, Kerala, India. She has organized several international conferences, is a fellow of the American Medical Society, and is a member of many international organizations. She has five publications to her name and has presented 25 papers.

Dr. Yang Weimin

Dr. Yang Weimin is the Taishan Scholar Professor of Quingdao University of Science and Technology in China. He is a full professor at the Beijing University of Chemical Technology and a fellow of many professional organizations. Professor Weimin has authored many papers in international peer-reviewed journals in the area of polymer processing. He has been contributed to a number of books as author and editor and acts as a reviewer to many international journals. In addition, he is a consultant to many polymer equipment manufacturers. He has also received numerous award for his work in polymer processing.

Contents

List of Contributors

Ali H. Al-Mowali
Chemistry department, College of Science, Basra University, Basra-Iraq.

N. Angulakshmi
Department of Physics, Lady Doak College, Madurai 625 002, India.

M. Bijarimi
Faculty of Chemical and Natural Resources Engineering, Universiti Malaysia Pahang, Malaysia.

Yuri M. Boiko
A.F. Ioffe Physico-Technical Institute of the Russian Academy of Sciences, Laboratory of Physics of Strength, 26 Politekhnicheskaya Str., St. Petersburg 194021, Russia.

L. Cerdán
Instituto de Química Física "Rocasolano", CSIC, Serrano 119, 28006 Madrid, Spain.

A. Costela
Instituto de Química Física "Rocasolano", CSIC, Serrano 119, 28006 Madrid, Spain.

R. Nimma Elizabeth
Department of Physics, Lady Doak College, Madurai 625 002, India.

O. García
Instituto de Ciencia y Tecnología de Polímeros, CSIC, Juan de la Cierva 3, 28006 Madrid, Spain.

I. García-Moreno
Instituto de Química Física "Rocasolano", CSIC, Serrano 119, 28006 Madrid, Spain.

P.N. Gupta
Physics Department, Banaras Hindu University, Varanasi-221005 (UP), India.

W. B. Gurnule
Department of Chemistry, Jagat Arts, Commerce and Indiraben Hariharbhai Patel Science College, Goregaon-440 018 Gondia, Maharashtra, India.

Hussain R. Hassan
Petrochemical Complex No.1, Basra-Iraq.

Moayad N. Khalaf
Chemistry department, College of Science, Basra University, Basra-Iraq.

B. Khemchandani
Department of Chemistry, Janta Vedic College, Baraut, 250611, India.

V. Martín
Instituto de Química Física "Rocasolano", CSIC, Serrano 119, 28006 Madrid, Spain.

K.S. Nahm
School of Chemical Engineering and Technology, Chonbuk National University, Chonju 561-756, S. Korea.

Vikas Nogriya
Department of Postgraduate Studies and Research in Physics and Electronics Rani Durgavti University, Jabalpur-482001 (M.P.) India.

M. E. Pérez-Ojeda
Instituto de Química Física "Rocasolano", CSIC, Serrano 119, 28006 Madrid, Spain.

G.K. Prajapati
Physics Department, Banaras Hindu University, Varanasi-221005 (UP), India.

S. S. Rahangdale
Department of Chemistry, Kamla Nehru Mahavidyalaya, Nagpur-440 009, Maharashtra, India.

Meera Ramrakhiani
Department of Postgraduate Studies and Research in Physics and Electronics Rani Durgavti University, Jabalpur-482001 (M.P.) India.

R. Roshan
Physics Department, Banaras Hindu University, Varanasi-221005 (UP), India.

Azemi Samsuri
Faculty of Applied Sciences, University Technology MARA, Malaysia.

Muthusamy Sarojadevi
Department of Chemistry, Carleton University, 1125 Colonel, by Drive, Ottawa, Ontario K1S 5B6, Canada and Department of Chemistry, Anna University, Chennai 600 025, India.

R. Sastre
Instituto de Ciencia y Tecnología de Polímeros, CSIC, Juan de la Cierva 3, 28006 Madrid, Spain.

Palaniappan Selvakumar
Department of Chemistry, Carleton University, 1125 Colonel by Drive, Ottawa, Ontario K1S 5B6, Canada and Department of Chemistry, Anna University, Chennai 600 025, India.

Pudupadi Sundararajan
Department of Chemistry, Carleton University, 1125 Colonel by Drive, Ottawa, Ontario K1S 5B6, Canada and Department of Chemistry, Anna University, Chennai 600 025, India.

V. Swaminathan
School of Materials Science and Engineering, 50 Nanyang Avenue 8, Singapore 639 798.

Sabu Thomas
Mahatma Gandhi University, Kottayam 686560, India.

H. S. Verma
Department of Chemistry, Janta Vedic College, Baraut, 250611, India.

Raed K. Zidan
Chemistry department, College of Science, Basra University., Basra-Iraq.

H., Zulkafli
Faculty of Chemical and Natural Resources Engineering, Universiti Malaysia Pahang, Malaysia.

List of Abbreviation

AFM	Atomic force microscopy
API	American petroleum Institute
ASE	Amplified spontaneous emission
BPTA	Benzophenonetetracarboxylic dianhydride
BTDA	Biphenyltetracarboxylic dianhydride
CB	Conduction band
DLS	Dynamic light scattering
DMF	Dimethylformamide
DMSO	Dimethyl sulfono oxide
DMSO	Dimethylsulphoxide
DSC	Differential scanning colorimeter
ED	Electron diffrection
EL	Electroluminescence
ELD	Electroluminescent devices
EMA	Effective mass approximation
6FDA	Hexafluroisopropylidene diphthalic anhydride
FT-IR	Fourier transformation infrared
FWHM	Full width at halh maximum
GPC	Gel permeation chromatography
HDPE	High density polyethylene
ILs	Ionic liquids
IR	Infra-red
LDPE	Low density polyethylene
MA	Maleic anhydride
MAPE	Maleated polyethylene
MMA	Methyl methacrylate
8MMAPOSS	Octapropyl methacryl-POSS
MW	Microwave
NCPEs	Nano composite polymer electrolytes
NMR	Nuclear magnetic resonance
NRFL	Non-resonant feedback lasing
OCP	Olefin based copolymer
ODA	Oxydianiline
Pas	Processing aids
PEO	Poly ethylene oxide

PET	Poly ethylene terephthalate
PFT	Power Fourier Transform
PL	Photoluminescence
PMA	Polymethaacrylate
PMDA	Pyromellitic dianhydride
PMMA	Poly methyl methacrylate
PMT	Photo multiplier tube
POSS	Polyhedral oligomeric silsesquioxanes
PS	Polystyrene
PVA	Poly vinyl alcohol
PVDF-HFP	Poly vinylidene fluoride-hexafluoro propylene
SAED	Selected area electron diffrection
SAIF	Sophisticated analytical instrumentation facility
SCPI	State Company for Petrochemical Industry
SEM	Scanning electron microscopy
SIR	Single internal reflection
SPE	Solid polymer electrolytes
SSDL	Solid-state dye lasers
TEM	Transmission electron microscopy
TMS	Tetramethylsilane
TPV	Thermoplastic vulcanized
VII	Viscosity index improvers
VM	Viscosity modifier
XRD	X ray diffrection
ZnS/PVA	Zinc sulfide/polyvinyl alcohol

Preface

Material Science as a whole has seen extra ordinary development, research interest and investment by industry in recent decades. Within this broad field, Polymer Science and Technology has in particular witnessed major strides. Indeed, polymers have virtually moulded the modern world and transformed the quality of life in innumerable areas of human activity. They have added new dimensions to standards of life and inexpensive product development. From transportation to communication, entertainment to health care, the world of polymers has touched them all. Polymer blends, composites, membranes, gels and polyelectrolytes occupy a unique position in the dynamic world of new materials. Polymer Science and Technology has attained its present status by means of exchange of ideas and the transfer of cutting edge technology through various platforms where academicians and industrialists have been brought together. Yet there is much more to conceive and achieve in this exciting field.

These papers are dealt with the importance of polymers in technology and some new inventions towards the field.

— Sabu Thomas, PhD

Chapter 1

Adhesion and Surface Glass Transition of Amorphous Polymers

Yuri M. Boiko

INTRODUCTION

The process of adhesion at the contact zone of two thick films of amorphous high-molecular-weight polymers with glassy bulk has been investigated by means of a lap-shear joint method. The kinetics and temperature dependence of strength σ at both symmetric and asymmetric polymer-polymer interfaces have been found to follow the repetition mechanism of chain diffusion and an Arrhenius-like behavior, respectively. A new method to measure the surface glass transition temperature ($T_g^{surface}$) of thick bulk polymer samples has been proposed. The values of $T_g^{surface}$ and of the glass transition temperature of the bulk (T_g^{bulk}) have been compared.

Over the two last decades, much attention has been paid to an increased molecular mobility at polymer surfaces and interfaces in comparison with that in the interior bulk regions (Akabori et al., 2003; Boiko et al., 1997, 2004; Fischer, 2002; Fu et al., 2005; Hyun et al., 2001; Kajiyama et al., 1995; Kawaguchi et al.,1998, 2003; Mansfield et al., 1991; Mayes, 1994; Prucker et al., 1998; Satomi et al., 1999; Zhang et al., 2000). Of particular interest is the question regarding the difference between the surface glass transition temperature $T_g^{surface}$ of the sample of an amorphous polymer and its glass transition temperature of the bulk T_g^{bulk} For example, a reduction in $T_g^{surface}$ with respect to T_g^{bulk} has been observed or predicted (Akabori et al., 2003; Fischer, 2002; Fu et al., 2005; Hyun et al., 2001; Kajiyama et al., 1995; Kawaguchi et al., 1998; Mayes, 1994; Prucker et al., 1998; Satomi et al., 1999). However, there is still the discussion in the literature as whether the reduction in $T_g^{surface}$ with respect to T_g^{bulk} is exclusively due to the presence of free surface or not (Boiko et al., 1997, 2004; Kawana et al., 2001; Sharp et al., 2003).

Most measurements of $T_g^{surface}$ have been performed on thin PS films (with a thickness δ of some tens to some hundreds of nm), while there is the lack of information concerning the $T_g^{surface}$ of the bulk thick samples (with δ of some hundreds of μm to some mm) of the engineering use, and for the polymers with other molecular architectures.

Hence, if the surface and interfacial properties of polymers are important for their application, it is important to know the difference, if any, between $T_g^{surface}$ and T_g^{bulk}. In

---

*Presented at the 2nd International Conference on Polymer Processing and Characterization (Kottayam, Kerala, India, January 15–17, 2010)

this connection, an important question arises concerning the measuring of $T_g^{surface}$ of the bulk polymer samples using a simple technique.

It has been shown earlier that the process of adhesion at the polymer-polymer interfaces of the samples with vitrified bulk is a diffusion process (Boiko et al., 1997, 2004). However, there is still uncertainty as by which kinetic unit of motion is controlled an elementary act of this process, in particular, at an asymmetric polymer-polymer interface experiencing an unfavorable (repulsive) thermodynamic segment-segment interaction.

Thus, the purpose of this study is threefold: (1) to propose a simple technique for measuring of $T_g^{surface}$ and to find out the difference, if any, between the $T_g^{surface}$ and T_g^{bulk} of the amorphous polymers; (2) to answer to the question as whether an increased segmental mobility in polymers with vitrified bulk exists on the free surface only; and (3) to find out the size of the kinetic unit of motion controlling an elementary act of the process of adhesion at the interface of polymers with vitrified bulk.

EXPERIMENTAL

Polymers

Polymers with four different types of chain architecture were selected: polystyrene (PS), poly (2,6-dimethyl-1,4-phenylene oxide) (PPO), poly(methyl methacrylate) (PMMA) and poly(ethylene terephthalate) (PET). The repeat units of these polymers are shown in Figure 1. As seen, PS and PMMA have only carbon atoms in the chain backbone. The PPO and PET have both carbon and oxygen atoms, and also have aromatic rings in the chain backbone due to which the chain becomes more rigid. However, PET has flexible -CH_2-CH_2- segments. The overall effect of these three factors of chain architecture results in the close values of the T_g^{bulk} for PET (81°C), PS (97–106°C), and PMMA (109°C), and in a much higher value of the T_g^{bulk} for PPO of 216°C (see Table 1). The polymers investigated were high-molecular-weight polymer since they had the molecular weight larger than the entanglement molecular weight (18,000 (Composto et al., 1992; Green et al., 1986), 8,800 (Willett et al., 1993), 3,400 (Green et al., 1986; Composto et al., 1992), 3,200 g/mol (Aharoni, 1978) for PS, PMMA, PPO, and PET, respectively) (see Table 1).

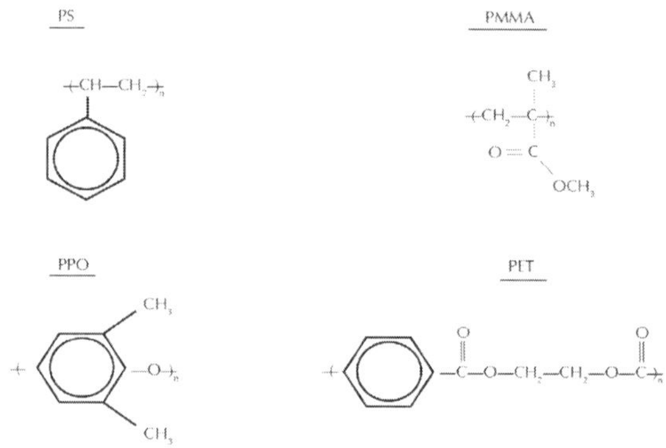

Figure 1. Chemical structure of the repeat units of polymers investigated.

Table 1. Some characteristics of polymers investigated.

Polymer	M_w, kg/mol	M_n, kg/mol	T_g^{bulk} (DSC)*, °C
PS	103–1,111	75–967	97–106
PPO	44	23	216
PMMA	87	43.5	109
PET	15	–	81

* Middle point of the corresponding heat capacity jump measured at a heating rate of 10°C/min on a Perkin-Elmer calorimeter.

Samples

The samples with a thickness δ of 100 μm having smooth surfaces were prepared by extrusion, or by compression molding between the smooth surfaces of silicone glass. The samples were considered as bulky monolithic samples, since the thickness of the surface layer, taken as the size of a statistical coil of an unperturbed chain (two radii of gyration R_g), was smaller than 0.1% of the sample thickness. Thus, the chain confinement effects relevant to ultrathin polymer films with $\delta < 2R_g$ (Dalnoki-Veress et al., 2001; Keddie et al., 1994a, 1994b; Mattsson et al., 2000) were excluded in such samples.

Healing Procedure

The samples were bonded in the lap-shear joint geometry (the contact area was 5 mm × 5 mm) at a small contact pressure of 0.2 MPa required to establish physical and thermal contact. This pressure was too small to give rise to viscous flow. Healing (or bonding) time t_h varied from 10 min to 24 hr.

Fracture Tests

As bonded adhesive joints (see Figure 2) were shear-fractured in tension on an Instron tensile tester at ambient temperature and at a crosshead speed of 5 mm/min. Lap-shear strength σ was calculated as fracture load divided by the contact area, averaged from at least 10 measurements. More details of the experimental procedures can be found elsewhere (Boiko et al., 1997, 2004).

RESULTS AND DISCUSSION

For investigating the kinetics of the development of adhesive strength σ for the samples with vitrified bulk, σ was measured as a function of healing time t_h for several polymer-polymer interfaces. A typical example of such dependence is shown in Figure 3 for a symmetric PS-PS interface. As seen, this curve is non-linear, which is expected for the diffusion-controlled processes in polymers; bonding is observed at T_h that is lower than T_g^{bulk} by 33°C. At the present time, it is generally accepted that the evolution of σ is directly proportional to the depth of penetration according to the Wool's minor chain repetition model (Kim et al., 1983; Wool et al., 1981,1995), that is to the one-fourth power of t_h, for the viscoelastic state of the bulk, which implies the viscoelastic state of the surface as well (the repetition mechanism of chain diffusion assumes a snake-like chain displacement along its contour in a highly entangled medium

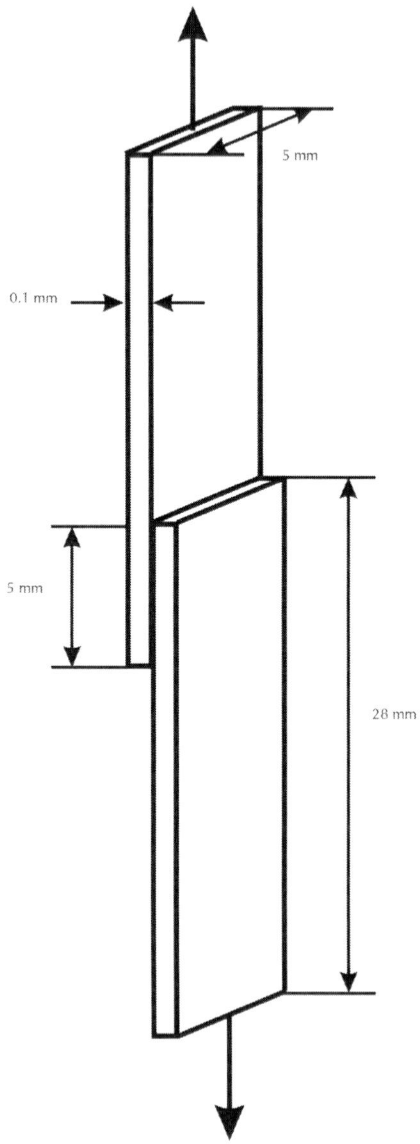

Figure 2. Geometry of lap-shear joint used in this work submitted to tensile loading.

(De Gennes, 1971). In order to find out whether this scaling law is valid for the contact zone of the samples with vitrified bulk, the curves σ versus $t_h^{1/4}$ were obtained for the PS-PS interface at several healing temperatures $T_h < T_g^{bulk}$ (see Figure 4(a)). As follows from Figure 4(a), a satisfactory linearity (obtained by a least-squares analysis; the same procedure was applied for the analysis of the experimental data presented in Figures 4(b), 5(a), 5(b), 6(a) and 6(b) below) in the coordinates $\sigma - t_h^{1/4}$ is observed at all the T_h's investigated. This behavior indicates that the process of adhesion is controlled by the repetition mechanism of chain diffusion.

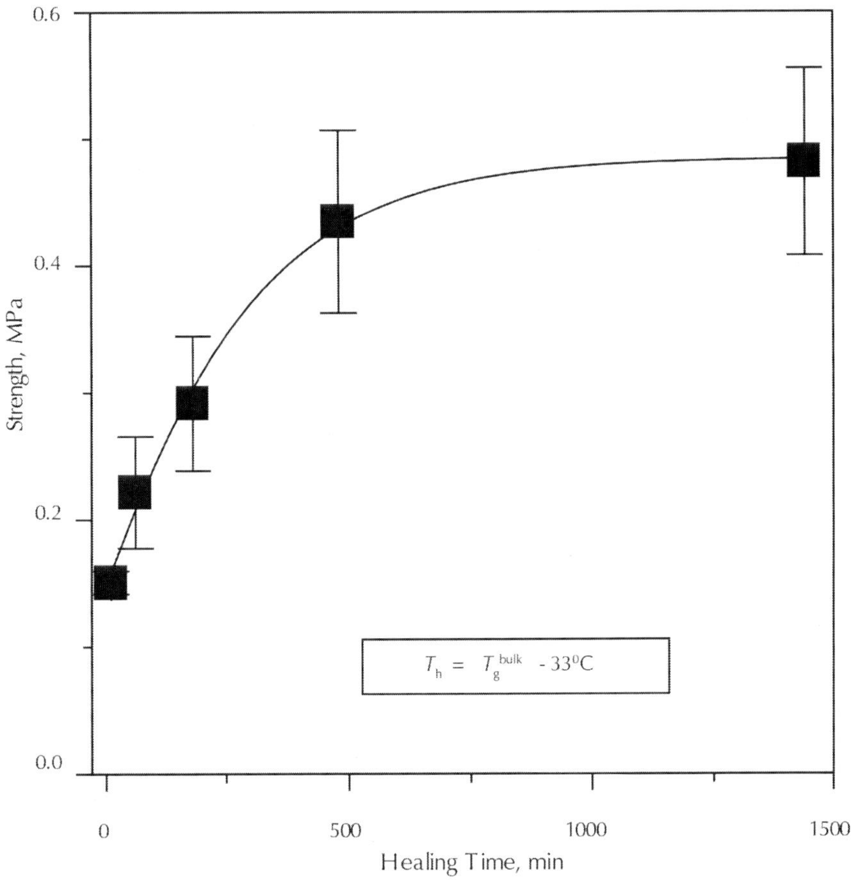

Figure 3. Lap-shear strength as a function of healing time at a healing temperature of $T_h = T_g^{bulk} - 33°C$ for a PS-PS interface with $M_w = 215{,}000$ g/mol; $T_g^{bulk} = 97°C$. A solid line is drawn as a guide to the eye.

As in the case of any diffusion process, one should expect an Arrhenius-like behavior for the repetition diffusion coefficient D_{rept} in the following form:

$$D_{rept} = D_0 \exp(-E_a/RT_h) \quad (1)$$

where D_0 is the pre-exponential factor, E_a is the activation energy, and R is the universal gas constant. Taking into consideration that (Kim et al., 1983; Wool et al., 1981, 1995).

$$D_{rept}^{1/4} \sim d\sigma / dt_h^{1/4} \quad (2)$$

and, combining Equations (1) and (2), we obtain Equations (3) and (4),

$$d\sigma / dt_h^{1/4} = c[D_0 \exp(-E_a/RT_h)]\, dt_h^{1/4} \tag{3}$$

where c is a constant,

$$\ln(d\sigma / dt_h^{1/4}) = \ln c + (1/4)\ln D_0 - E_a/4RT_h \tag{4}$$

that gives us, finally, Equation (5):

$$E_a(D) = 4R\,[\Delta \ln(d\sigma / dt_h^{1/4}) / \Delta(1/T_h)] \tag{5}$$

In order to investigate the correspondence of the data presented in Figure 3(a) to Equation (5), the logarithm of the slope to the curve $\sigma - t_h^{1/4}$ is plotted as a function of $1/T_h$ in Figure 4(b). As seen, this plot has a linear shape that gives us an opportunity to calculate, using Equation (5), the value of $E_a(D) = 250$ kJ/mol. This value is rather high and, hence, it is apparent. Nevertheless, it will be used below for a comparative analysis.

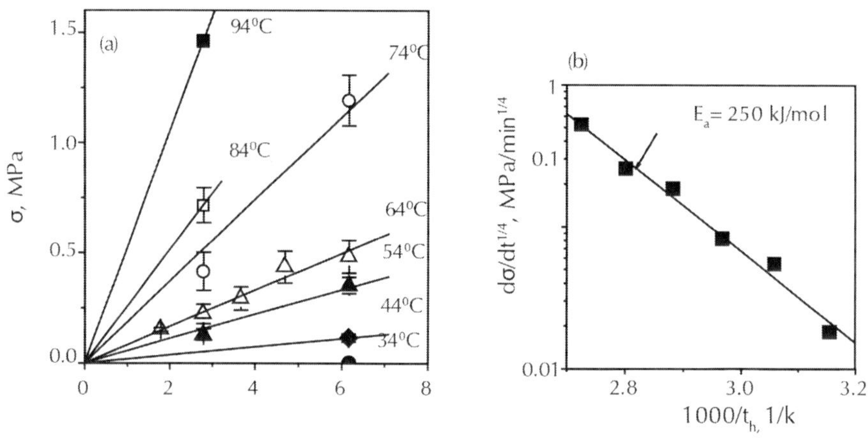

Figure 4. (a) Lap-shear strength as a function of healing time to the one-fourth power at several healing temperatures below T_g^{bulk}, and (b) the logarithm of slope $\sigma - t^{1/4}$ as a function of reciprocal healing temperature, for a symmetric PS-PS interface with $M_w = 215{,}000$ g/mol.

Having applied the same procedure to the experimental data for a symmetric PMMA-PMMA (at $T_h < T_g^{bulk}$) and an asymmetric PS-PMMA interfaces (see Figures 5(a) and 6(a), respectively), one can, first, construct the plots log $(d\sigma/dt_h^{1/4})$ versus $1/T_h$ (see Figures 5(b) and 6(b)) and, basing on their satisfactory linearity, to calculate the values of $E_a(D)$: 160 and 170 kJ/mol for the PMMA-PMMA and PS-PMMA interfaces, respectively. It is interesting that the strength at an incompatible PS-PMMA interface increases directly proportionally to $t_h^{1/4}$ (see Figure 6(a)), as in the case of the compatible PS-PS and PMMA-PMMA interfaces. It indicates that, despite the repulsive interaction between the segments of PS and PMMA (Wool, 1995), the repetition process controls the strength evolution at the incompatible PS-PMMA interface as well. In other words, the thermodynamics of intermolecular interaction (neutral in the case of the PS-PS and PMMA-PMMA interfaces and unfavorable in the case of the PS-PMMA interface) does not principally change the kinetics of strength development at the interfaces of polymers with glassy bulk.

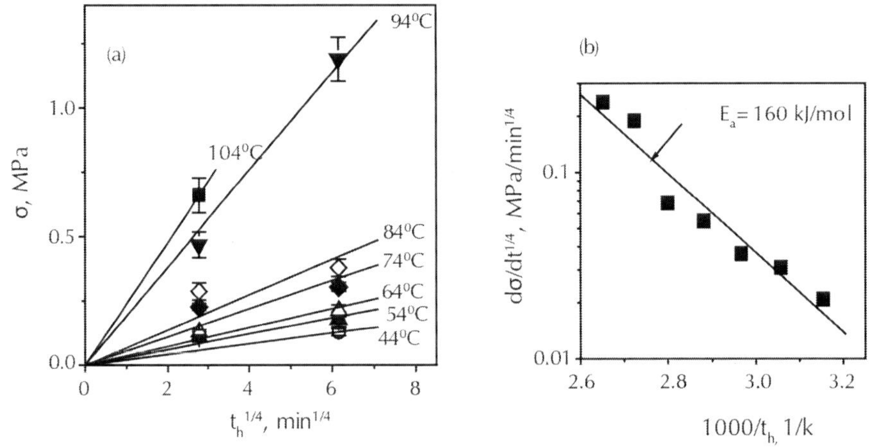

Figure 5. (a) Lap-shear strength as a function of healing time to the one-fourth power at several healing temperatures below T_g^{bulk}, and (b) the logarithm of slope $\sigma - t^{1/4}$ as a function of reciprocal healing temperature, for a symmetric PMMA-PMMA interface with M_w = 87,000 g/mol.

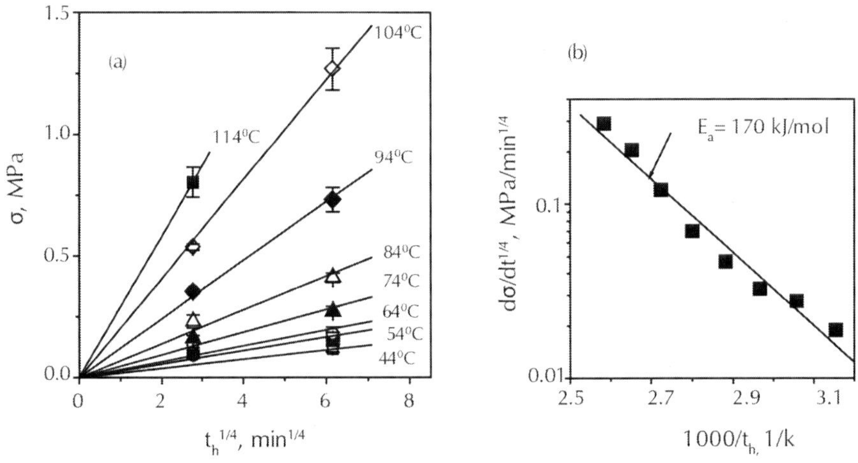

Figure 6. (a) Lap-shear strength as a function of healing time to the one-fourth power at several healing temperatures, and (b) the logarithm of slope $\sigma - t^{1/4}$ as a function of reciprocal healing temperature, for an asymmetric PS-PMMA interface (the values of M_w are 215,000 and 87,000 g/mol for PS and PMMA, respectively).

Thus, it may be concluded that the process of adhesion of the samples with vitrified bulk is a diffusion process. Despite the glassy state of the bulk wherein the segmental motion in the chain backbone is frozen, there is a temperature interval where the contact zone (and, therefore, the free surface prior to contact) is in the viscoelastic state.

The two values of E_a (D) of 160 and 170 kJ/mol for the PMMA-PMMA and PS-PMMA interfaces are close, and they are smaller than the value of E_a $(D) = 250$ kJ/mol for the PS-PS interface. This behavior indicates that the process of inter diffusion at the PS-PMMA interface is controlled by a slow component, PMMA. Indeed, on the one hand, the values of σ at the PS-PS interface are larger than those at the PMMA-PMMA interface, and, on the other hand, the values of σ at PMMA-PMMA and PS-PMMA interfaces are close (see Figures 4(a), 5(a) and 6(a)).

One of the main reasons of the existence of the viscoelastic layer on the surface of the sample with vitrified bulk may be a decrease in both mass and entanglement density at free polymer surfaces (Brown et al., 1996; Mansfield et al., 1991) favoring more intensive molecular motion in comparison with that in the bulk. If this is true, one should expect a decrease in the energetic barrier that should be overcome for the activation a long-range mode of segmental motions in polymers (α-relaxation) at the surface and interface in comparison with that in the bulk. For investigating this hypothesis, let us compare the values of E_a (D) obtained in this work for the glassy state of the bulk with those available in the literature for the processes of diffusion and α-relaxation, E_a (α). Since PS turned out to be one of the polymers investigated in great details in this respect, the analysis will be performed for PS, and the corresponding data are collected in Table 2.

Table 2. Activation energies of the processes of diffusion and α-relaxation in PS.

M_w, kg/mol	Analyzed Zone	E_a, kJ/mol (process)	Interval of Measurements, °C
103–1,111	PS-PS Interface	250–300 (diffusion), present work	44–93
29–1,082	PS-PS Interface	210–360 (diffusion) [5, 30-32]	80–170
140–1,460	PS Surface	210–270 (a-relaxation) [6-8]	20–120
110	PS Bulk	560 (diffusion) [33]	108–120
4–1,000	PS Bulk	400–480 (a-relaxation) [34]	Surroundings of T_g^{bulk}

From Table 2 follows that, first, the values of E_a $(D) = 250–300$ kJ/mol obtained in this work are close to the values of E_a $(D) = 210–360$ kJ/mol obtained in (Guérin et al., 2003; Kawaguchi et al., 2003; Kline et al., 1988; Whitlow et al., 1991), both of them being independent of the molecular weight. It means that the kinetic unit of motion that controls an elementary act of this process is smaller than the entire chain. Moreover, these values of E_a (D) are close to the values of E_a for the α-relaxation on the free PS surface E_a $(\alpha)^{surface} = 210–270$ kJ/mol (Akabori et al., 2003; Fu et al., 2005; Kawaguchi et al. 1998). Hence, one may conclude that the process of repetition at the PS-PS interfaces is controlled by an elementary act of α-relaxation (conformational transitions in the chain backbone), and the persistence length of this kinetic unit is equal to statistical Kuhn's segment. Second, both the values of E_a (D) at the interface and of E_a $(\alpha)^{surface}$ are smaller than those of E_a (D) in the bulk (Jou, 1986) and E_a $(\alpha)^{bulk}$ (Bershtein et al., 1994) by a factor of 2. It indicates that the decrease in E_a at free polymer surfaces in comparison with the bulk and the retention of this effect at polymer interfaces is one of the key reasons of the existence of the increased molecular mobility. This behavior may be explained by a decreased number of segments-neighbors and an increased distance between them, both being provided by an increase in the free volume fraction at the surface.

Owing to the diffusion-controlled development of adhesive strength for the contacted samples with glassy bulk observed above, one may detect the lowest healing temperature T_h^{lowest} when autoadhesion still takes place due to some (though very limited) diffusion of chain segments across the interface. In its turn, the occurrence of diffusion is not feasible without the realization of the conformational (*gauche-trans*) transitions in the chain backbone, which is a long-range motion (the process of α-relaxation). Hence, T_h^{lowest} may be considered as the T_g at the interface or as a high limit of $T_g^{surface}$ (if $T_g^{surface}$ heightens upon contact).

Measuring of Surface Glass Transition Temperature

In order to find out as whether the lap-shear joint method may be used to measure $T_g^{surface}$, in Figure 7 are plotted the values of the lap-shear strength σ for four symmetric polymer-polymer interfaces as a function of healing temperature at a chosen t_h = 1 hr. As follows from Figure 7, the values of σ for all the interfaces investigated decrease monotonically with a reduction in T_h, approaching a very small level of σ (some hundredths of MPa) at T_h^{lowest} = 44°C (PS and PMMA), 64°C (PET), and 90°C (PPO), and, finally, dropping to σ →0. It means that diffusion did not occur at $T_h < T_h^{lowest}$. In terms of fracture stress, it is not easy to say as whether the small values of σ = 0.02–0.05 MPa developed at T_h^{lowest} are due to some interdiffusion. However, it terms of fracture energy (the total work of deformation A accomplished in the course of fracture of the unit of the contact area), the chain diffusion may be easily recognized once the value of A becomes larger than the thermodynamic work of autoadhesion W_a. Actually, W_a characterizes the physical attraction of the contacting surfaces via weak intermolecular van-der-Waals' forces only, without involving any interdiffusion (Wool, 1995).

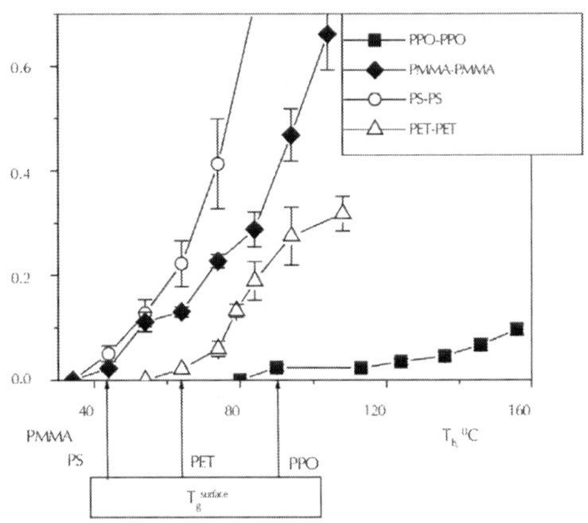

Figure 7. Lap-shear strength as a function of healing temperature at symmetric PS-PS, PMMA--PMMA, PPO-PPO, and PET-PET interfaces; healing time is 1 hr.

It has been shown earlier (Boiko, 2000) that small values of $\sigma \geq 0.02$ MPa correspond to values of $A \geq 0.12$ J/m^2 for the interfaces investigated that are still larger than the values of $W_a < 0.09$ J/m^2 (Van Krevelen, 1997). Since W_a characterizes the physical attraction of the contacting surfaces via weak intermolecular van-der-Waals' forces only, without involving any interdiffusion, one may conclude that some diffusion of chain segments across the interface still takes place at T_h^{lowest}. Therefore, the inability of bonding between two like polymer solids at $T_h < T_h^{lowest}$ is caused by freezing of the segmental motions at the interface, and, hence, $T_h^{lowest} \approx T_g^{surface}$. The values of $T_g^{surface}$ for the polymers investigated measured by the approach proposed are indicated with arrows on the abscissa of Figure 7 (44°C for PS and PMMA, 64°C for amPET, and 90°C for PPO and crPET). Having determined the values of $T_g^{surface}$, we can now compare them with the values of T_g^{bulk}.

Such a comparison is presented in Table 3. From Table 3 follows, first, that $T_g^{surface}$ is lower than T_g^{bulk} by some tens degrees of centigrade for all the polymers investigated, of even by more than 100°C in the case of PPO. Therefore, it may be concluded that the lap-shear joint method is a useful technique for measuring the surface glass transition in the amorphous polymers. Second, the reduction in $T_g^{surface}$ becomes larger with an increase in T_g^{bulk}, that is with an increase in the intensity of the intermolecular interaction in the bulk. This behavior may be explained by a more notable weakening of the intermolecular interaction on the surface (a more notable decrease in the energetic barrier that should be overcome in order to activate the segmental motion on the surface in comparison with the bulk) of the polymers with higher T_g^{bulk}.

Table 3. Values of T_g^{bulk} and of the Effect of the Reduction in $T_g^{surface}$ with Respect to T_g^{bulk} for Amorphous Polymers Investigated

Polymer	$T_g^{surface}$, °C	T_g^{bulk}, °C	$T_g^{surface} - T_g^{bulk}$, °C
PPO	90	216	–126
PMMA	44	109	–65
PS	44	97	–53
PET	64	81	–17

Actually, the value of the reduction in $T_g^{surface}$ in regard to T_g^{bulk} correlates with the difference between the values of the activation energy of interdiffusion E_a (D) determined by the procedure described above and those of the process of α-relaxation in the bulk $E_a(\alpha)^{bulk}$ available in the literature (Bershtein et al., 1994; Hwang, 1997; Linghu et al., 2000) (see Figure 8) these two characteristics are connected by a linear relationship. The data of Figure 8 indicate that the larger is the difference $[E_a(\alpha)^{bulk} - E_a(D)]$ the smaller is the value of $(T_g^{surface} - T_g^{bulk})$, that is the more notable effect of the reduction in $T_g^{surface}$ with respet to T_g^{bulk} is observed.

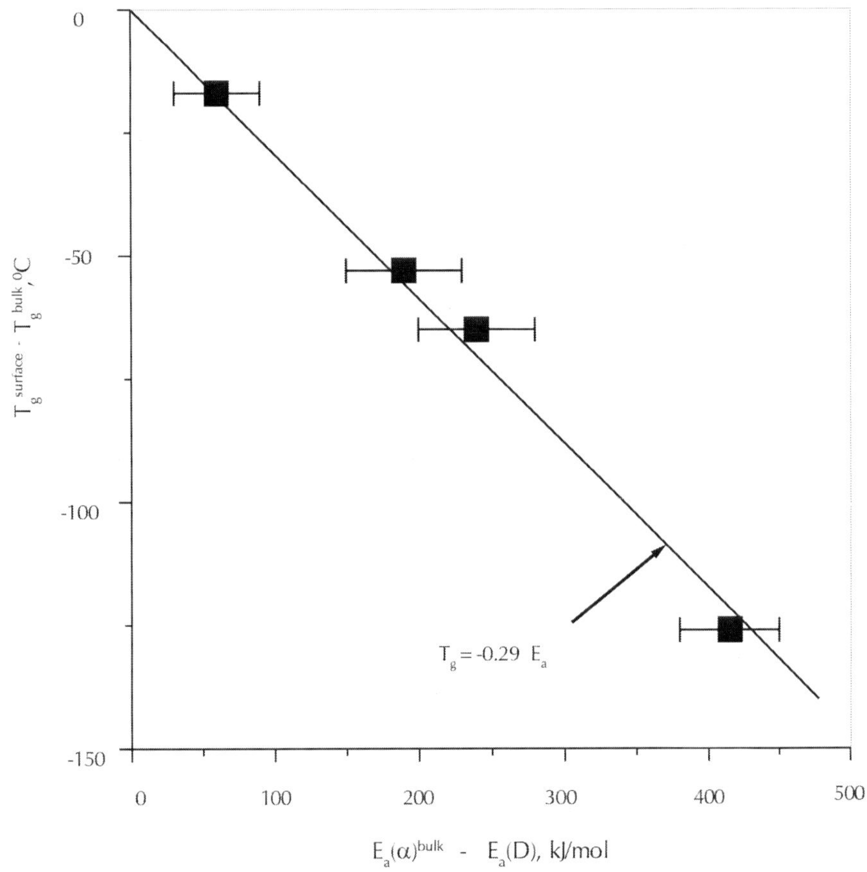

Figure 8. Effect of a reduction in $T_g^{surface}$ with respect to T_g^{bulk} (ΔT_g) as a function of the difference between E_a (α-relaxation) in the bulk [34, 37, 38] and E_a (diffusion) at the interface of polymers with vitrified bulk (ΔE_a).

Finally, the occurrence of the diffusion across the contact zone of two polymer samples with vitrified bulk indicates that the long-range segmental motions are realized in such samples not exclusively on free surfaces, as was suggested in (Kawana et al., 2001; Sharp et al., 2003) but in the contact layers as well.

CONCLUSION

It has been elucidated that the process of adhesion between the two contacting polymer samples with vitrified bulk is controlled by the repetition mechanism of chain diffusion independently of whether the interface experiences neutral or unfavorable thermodynamic segment-segment interaction; the kinetic unit of motion of an elementary act of this process is statistical Kuhn's segment. It has been shown that the lap-shear joint mechanical adhesive method is an appropriate technique for measuring of the $T_g^{surface}$

of the thick polymer films. It has been found that the $T_g^{surface}$ of an amorphous polymer may be decreased by some tens, or even by 100, degrees of Kelvin with respect to the T_g^{bulk} of the polymer. This effect persists at the contact zone of such samples, and it becomes more notable with an increase in the intermolecular interaction in the bulk. The existence of this effect correlates well with the decrease in the activation energies of the conformational transitions at surfaces and interfaces in comparison with those in the bulk.

KEYWORDS

- **Adhesion**
- **Amorphous polymers**
- **Interdiffusion**
- **Surface glass transition**
- **Vitrified bulk**

Chapter 2

Chelating Ion-exchange Properties of Copolymer Resins

W. B. Gurnule and S. S. Rahangdale

INTRODUCTION

Copolymer resin (p-CAF) was synthesized by the condensation of p-cresol and adipamide with formaldehyde in the presence of hydrochloric acid as catalyst and using varied molar ratios of reacting monomers. A composition of the copolymers has been determined on the basis of their elemental analysis. The number average molecular weight of resins was determined by conductometric titration in non-aqueous medium. The copolymer resins were characterized by viscometric measurements in dimethylsulphoxide (DMSO), UV-visible absorption spectra in non-aqueous medium, infrared (IR) spectra, and nuclear magnetic resonance (NMR) spectra. The morphology of the copolymers was studied by scanning electron microscopy (SEM).

So, far no resin based on p-cresol-adipamide-formaldehyde in DMF medium has been reported for the quantitative removal and separation of transition and post transition metal ions. As industrial effluents are often rich in transition and post transition metal ions, removal of these metals is an important industrial task. The work described in the present communication deals with the synthesis, characterization and the systematic studies of various ion-exchange properties of the resin.

Ion-exchange resins are produced and commercialized in a wide range of formulations with different characteristics, and have now a large practical applicability in various industrial processes, such as chemical, nuclear chemistry for treatment of liquid waste, pharmaceutical, food industry, and so on (Lokhande et al., 2006, 2007, 2008). For their versatile properties, the copolymer resins are used in the ion-exchange area and in the heterogeneous catalysis field (Lokhande et al., 2006, 2007).

Organic ion-exchange resins are the most important class of ion-exchangers. Several workers have investigated the role played by the polymer structures on ion-exchange equilibria and physical properties of the system. DeGeiso et al., (1962) developed the evaluation procedure for determining the selectivity of chelating polymer prepared from salicylic acid and formaldehyde. Gregore et al., (1962) reported the properties of exchangers containing chelate groups derived from anthranillic acid, resorcinol and formaldehyde.

Ion-exchange resins have been employed for many new inorganic and organic separations. Routine use of ion-exchange resins have been reported by Spedding et al., (1951, 1952). Recently Aly et al., (1999) have prepared ion-exchangers to remove Hepatitis A virus from drinking water. Chelating resin containing oxine groups are synthesized either from oxine, formaldehyde and resorcinol by condensation (Parrish,

1955; Von Lillin, 1954) or by diazotization of poly (aminostyrene) resin followed by coupling to oxine (Roy et al., 2004). A cross linked styrene/meleic acid chelating matrix has been reported for its higher ability to removed the metal ions such as Cr^{2+}, Fe^{3+}, Ni^{2+}, Cu^{2+}, and Pb^{2+} (Rivas et al., 2002). Acidic polymers such as poly (methacrylic acid) and poly (acrylic acid) have the tendency of removed the metal ions like Ag^{2+}, Cu^{2+}, Co^{2+}, Ni^{2+}, and Cr^{2+} at different pH and polymer metal ion ratios (Rivas et al., 2002). Salicylic acid melamine with formaldehyde terpolymer found to have higher selectivity for Fe^{3+}, Cu^{2+}, and Ni^{2+} ions then for Co^{2+}, Zn^{2+}, Cd^{2+}, and Pb^{2+} ions (Gurnule et al., 2002). Resin synthesized by the condensation of a mixture of phenol or hydroxybenzoic acid with formaldehyde and various amines have also been reported (Gurnule et al., 2003a). The metal ion uptake increases with increasing mole proportions of the terpolymer synthesized from substituted benzoic acid (Burkanudeen et al., 2003). The *o*-nitrophenol and thiourea with paraformaldehyde terpolymer was identified as an excellent ion-exchanger for Zn^{2+} and Co^{2+} ions (Burkanudeen et al., 2002). Salicylic acid-formaldehyde-resorcinol resin has been synthesized and explored its use for the removal and separation of heavy metal ions from their binary mixture (Shah et al., 2006). The 8-hydroxyquinoline-formaldehyde-catechol copolymer found to have lower moisture content indicating the high degree of cross linking in the resin (Shah et al., 2008). So far no resin based on p-cresol, adipamide, and formaldehyde in acid media has been synthesized for the quantitative separation of transition metal ions. The present chapter describes the development of a novel ion-exchange resin suitable for the desalination of waste water, which is high in Fe^{3+}, Cu^{2+}, and Ni^{2+} ions, to meet effluent discharge specifications. Ion-exchange column of p-CAF copolymer resin can be use for removal of Fe^{3+}, Cu^{2+}, and Ni^{2+} metal ions as well as suspended solid in waste water. It can also be used in the removal of Fe^{3+} from boiler water in industries (Feng et al., 2000). The resin can also be use for the removal of Fe^{3+} and Zn^{2+} from brass. There are many useful reports on ion-exchange separation methods in chemical process (Shah et al., 2001). Some commercially available ion-exchange resins are given Table 1.

Table 1. Commercially available ion-exchange resins.

Trade Name	Functional Group	Polymer Matrix	Ion-exchange Capacity (mmol.g^{-1})
Amberlite IR-120	$-C_6H_4SO_3H$	Polystyrene	5.0-5.2
Duolite C-3	$-CH_2SO_3H$	Phenolic	2.8-3.0
Amberlite IRC-50	-COOH	Methacrylic	9.5
Duolite ES-63	$-OP(O)(OH)_2$	Polystyrene	6.6
Zeocarb-226	-COOH	Acrylic	10.00
Dowex-1	$-N(CH_3)_3Cl$	Polystyrene	3.5
Amberlite IRA-45	$-NR_2$, -NHR, $-NH_2$	Polystyrene	5.6
Dowex-3	$-NR_3$, -NHR, $-NH_2$	Polystyrene	5.8
Allassion A WB-3	$-NR_2$, $-N^+R_3$	Epoxy-Amine	8.2

EXPERIMENTAL

Materials

The chemicals used were all of A.R. or chemically pure grade and are procured from the market. The cost of material synthesis is around 12,000 in Indian Rupees.

Synthesis of p-CAF Copolymer Resin

The p-CAF-I copolymer resin was prepared by condensing p-cresol (1.08 g, 0.1 mol) and adipamide (0.88 g, 0.1 mol) with formaldehyde (7.4 ml of 37% solution, 0.2 mol) in the presence of 2M HCl (200 ml) as a catalyst at 130°C in an oil bath for 5 hr with occasional shaking to ensure thorough mixing. The solid resinous product obtained was removed immediately from the flask. It was washed with cold water, dried and powdered. The powder was repeatedly washed with hot water to remove unreaced monomers. Then it was extracted with diethyl ether to remove excess of p-cresol-formaldehyde copolymer which might be present along with p-CAF copolymer resin. The purified copolymer resin was finely ground and kept in a vacuum over silica gel. The yield of the copolymer resin was found to be 80%.

Similarly, the other copolymer resins, p-CAF-II, p-CAF-III, and p-CAF-IV were synthesized by varying the molar proportion of the starting materials that is, p-cresol, oxamide and formaldehyde in the ratios 2:1:3, 3:1:4, and 4:1:5 respectively. The details of reaction and elemental analysis are depicted in Table 2.

Characterization of the Copolymers

The viscosities were determined by using TuanFuoss viscometer at six different concentrations ranging from 1.00 to 0.031% of copolymer in DMSO at 30°C. The intrinsic viscosity [η] was calculated by relevant plots of the Huggins' equation and Kraemmer's equation similar to earlier coworkers (Gurnule et al., 2003a; Shah et al., 2006).

The number average molecular weights ($\overline{Mn}$) were determined by conductometric titration in non-aqueous medium such as DMSO using ethanolic KOH as a titrant. Form the graphs of specific conductance against milliequivalents of base, first and last break were noted from which the degree of polymerization ($\overline{DP}$) and the number average molecular weight ($\overline{Mn}$) have been calculated for each copolymer resins.

Electron absorption spectra of all copolymer resins in DMSO (spectroscopic grade) were recorded on Shimadzu double beam spectrophotometer in the range of 200–850 nm at Sophisticated Analytical Instrumentation Facility (SAIF), Punjab University, Chandigarh, IR spectra of four p-COF copolymer resins were recorded on Perkin-Elmer-983 spectrophotometer in KBr pallets in the wave number region of 4000–400 cm^{-1} at SAIF, Punjab University, Chandigarh, NMR spectra of newly synthesized copolymer resins have been scanned on Bruker Advanced 400 NMR spectrometer using DMSO-d_6 at SAIF, Punjab University, Chandigarh. The SEM has been scanned by FEI-Philips XL-30 electron microscope.

Ion-exchange Properties

The ion-exchange properties of the p-CAF copolymer resins were determined by the batch equilibrium method (Gupta et al., 2008). The ion-exchange properties of all the

Table 2. Synthesis and physical data of p-CAF copolymer resins.

Copolymer	Reactants			Catalyst 2M HCl (ml)	Yield (%)	Colour	Melting point (K)	Elemental					
	p-cresol (mol)	Adipamide (mol)	Formaldehyde (mol)					C		H		N	
								Cal.	Found	Cal.	Found	Cal.	found
p-CAF-I	0.1	0.1	0.2	200	80	Yellow	388	65.21	66.34	6.24	6.10	12.14	12.13
p-CAF-II	0.2	0.1	0.3	200	75	Yellow	391	69.69	68.45	5.07	5.15	7.07	7.28
p-CAF-III	0.3	0.1	0.4	200	80	Yellow	394	72.09	73.91	6.97	6.12	5.42	5.35
p-CAF-IV	0.4	0.1	0.5	200	80	Yellow	398	73.35	71.72	6.89	6.35	4.38	4.43

four resins have been studied. However, only the data for the p-CAF-I copolymer resin has been presented in this chapter.

Determination of Metal Uptake in the Presence of Various Electrolytes and Different Concentration

The copolymer sample (25 mg) was suspended in an electrolyte solution (25 ml) of known concentration. The pH of the suspension was adjusted to the required value by using either 0.1 M HNO_3 or 0.1 M NaOH. The suspension was stirred for 24 hr at 30°C. To this suspension 2 ml of 0.1 M solution of the metal ion was added and the pH was adjusted to the required value. The mixture was again stirred at 30°C for 24 hr. The polymer was then filtered off and washed with distilled water. The filtrate and the washing were collected and then the amount of metal ion was estimated by titrating against standard EDTA (ethylene diamine tetra-acetic acid) at the same pH (experimental reading). The same titration has been carried out without polymer (blank reading). The amount of metal ion uptake of the polymer was calculated from the difference between a blank experiment without polymer and the reading in the actual experiments. The experiment was repeated in the presence of several electrolytes (Gupta et al., 2008; Jadhao et al., 2005a). Metal ion, its pH range, buffer used, indicator used and color change are given in Table 3. The metal ion uptake can be determined as,

Metal ion adsorbed (uptake) by resin = (XY) Z mmol/g

where,

Z (ml) is the difference between actual experimental reading and blank reading.

X (mg) is metal ion in the 2ml 0.1M metal nitrate solution before uptake.

Y (mg) is metal ion in the 2ml 0.1M metal nitrate solution after uptake.

By using this equation the uptake of various metal ion by resin can be calculated and expressed in terms of milli equivalents per gram of the copolymer.

Table 3. Data of experimental procedure for direct EDTA titration.

Metal Ion	pH range	Buffer used	Indicator used	Colour change
Fe(III)	2-3	Dil.HNO_3/dil.NaOH	Variamine blue	Blue-Yellow
Cu(II)	9-10	Dil.HNO_3/dil.NaOH	Fast sulphone black-F	Purple-Green
Ni(II)	7-10	Aq.NH_3/NH_4Cl	Murexide	Yellow-Violet
Co(II)	6	Hexamine	Xylenol orange	Red-Yellow
Zn(II)	10	Aq.NH_3/NH_4Cl	Salochrom black	Wine Red-Blue
Cd(II)	5	Hexamine	Xylenol orange	Red-Yellow
Pb(II)	6	Hexamine	Xylenol orange	Red-Yellow

Estimation of Rate of Metal Ion Uptake as Function of Time

In order to estimate the time require to reach the state of equilibrium under the given experimental conditions, a series of experiments of the type described above were carried out, in which the metal ion taken up by the chelating resins was determined from time to time at 30°C (in the presence of 25ml of 1M $NaNO_3$ solution). It was assumed that, under the given conditions, the state of equilibrium was established within 24 hr (Gupta et al., 2008). The rate of metal uptake is expressed as percentage of the amount of metal ions taken up after a certain time related to that at the state of equilibrium and it can be defined by the following relationship.

The percent amount of metal ions taken up at different time is defined as.

$$\text{Percentage of amount of metal ion taken up at different time} = \frac{\text{Amount of metal ion adsorbed}}{\text{Amount of metal ion adsorbed at equilibrium}} \times 100$$

Percentage of metal ion adsorbed after 1 hr = (100X)/Y

where, X is mg of metal ion adsorbed after 1 hr and Y is mg of metal ion is adsorbed after 25 hr, then by using this expression, the amount of metal adsorbed by polymer after specific time intervals was calculated and expressed in terms of percentage metal ion adsorbed. This experiment was performed using 0.1 M metal nitrate solution of Fe^{3+}, Cu^{2+}, Ni^{2+}, Co^{2+}, Zn^{2+}, Cd^{2+}, and Pb^{2+}.

Evaluation of the Distribution of Metal Ions at Different pH

The distribution of each one of the seven metal ions that is, Cu (II), Ni (II), Co (II), Zn (II), Cd (II), Pb (II), and Fe (III) between the polymer phase and the aqueous phase was determined at 30°C and in the presence of 1M $NaNO_3$ solution. The experiments were carried out as described above at different pH values. The distribution ratio, D, is defined by the following relationship (Patel et al., 2004).

$$D = \frac{Amount\ of\ metal\ ion\ on\ resin}{Amount\ of\ metal\ ion\ in\ solution} \times \frac{Volume\ of\ solution\ (ml)}{Weight\ of\ resin\ (g)}$$

$$\text{Metal ion adsorbed (uptake) by the resin} = \left(\frac{ZX}{Y}\right)\frac{2}{0.025}$$

where Z is the difference between actual experiment reading and blank reading, C (g) is the amount of metal ion in 2 ml 0.1 M metal nitrate solution, and Y (g) of metal ion in 2 ml of metal nitrate solution after uptake.

RESULT AND DISCUSSIONS

The four new copolymer resins p-CAF were synthesized by condensing p-cresol and adipamide with formaldehyde in the presence of 2 M HCl as catalyst in an oil bath for 5 hr in the molar ratios of 1:1:2, 2:1:3, 3:1:4, and 4:1:5. All four p-CAF copolymer resins were found to be yellow in color. The copolymers are soluble in DMF, DMSO and are insoluble in almost all other organic solvents. The melting points of these resins were found to be in the range of 118–130°C. These resins were analyzed for carbon, hydrogen, and nitrogen content. The p-CAF-I copolymer which has been used in the present investigation was prepared by the reaction given in Figure 1. The details of synthesis of copolymer along with color, melting point, yield and elemental analysis are incorporated in Table 2.

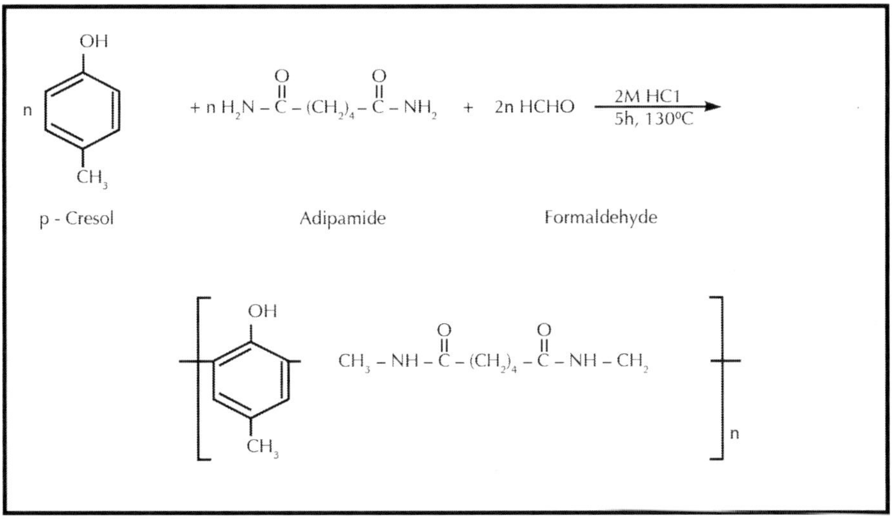

Scheme 1. Synthesis of representative p-CAF-I copolymer resin.

Figure 1. The SEM of p-CAF-I copolymer resin.

Viscometric Study

Viscometric measurements were carried out in DMSO solutions at 30°C using a Tuan–Fuoss (Gurnule et al., 2003a) viscometer fabricated in our research laboratory at a different concentrations ranging from 1.00 to 0.031%. Intrinsic viscosity (η) was calculated from relevant plots of Huggins' equation (1) and Kraemer's equation (2) (Shah et al., 2006).

$$\eta_{sp}/C = [\eta] + K_1[\eta]^2.C \quad (1)$$

$$\ln \eta_r/C = [\eta] - K_2[\eta]^2.C \quad (2)$$

where C = concentration in g/100 ml.

η_r = the ratio between viscosity of solution [η] and viscosity of the solvent $[\eta_0]$ is known as relative viscosity $\eta_r = \eta/\eta_0$

η_{sp} = this function has been derived from relative viscosity and given by

$\eta_{sp} = (\eta - \eta_0)/\eta_0 = \eta/\eta_0 - 1 = \eta_r - 1$

[η] = it is intrinsic viscosity obtained by extrapolating a plot of η_{sp}/C or in η_r/C against concentration. $[\eta] = \lim_{C\to 0} (\eta_{sp}/C)$. The intrinsic viscosity is characteristics parameter of a polymer.

According to above relations, the plots of η_{sp}/C and ln η_r/C against concentration was linear with slopes K_1 and K_2 respectively (Figure 2). Intercepts on the viscosity function axis give [η] value in both plots. The calculated values of the constant K_1 and K_2 (Table 2) in most cases satisfy the relation $K_1 + K_2 = 0.5$ favorably (Gupta et al., 2008). It was observed that copolymer having higher average molecular weight ($\overline{Mn}$) shows a higher value of intrinsic viscosity [η].

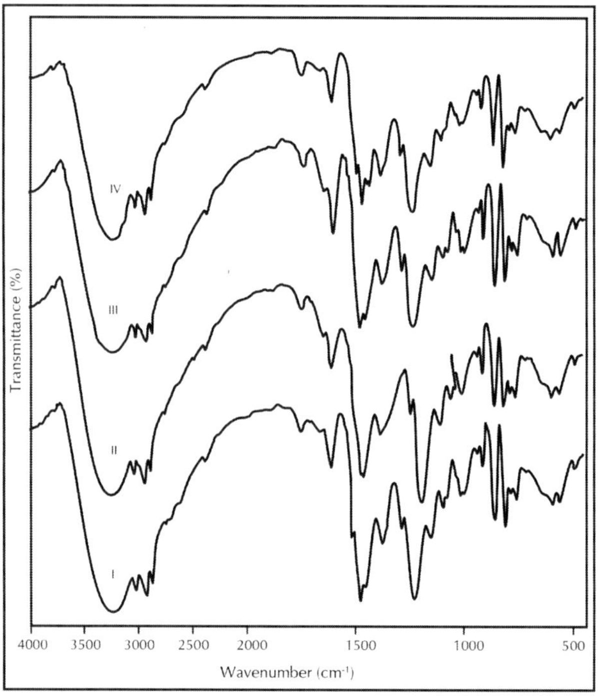

Figure 2. Infrared spectra of p-CAF copolymer resins (1) p-CAF-I, (2) p-CAF-II, (3) p-CAF-III, and (4) p-CAF-IV.

Average Molecular Weight

The number average molecular weight $(\overline{Mn})$ of these copolymers has been determined by conductometric titration method in non-aqueous medium and using standard potassium hydroxide (0.05 M) in absolute ethanol as a titrant. The results are depicted in Table 4. The specific conductance was plotted against milli equivalents of ethanolic KOH required for neutralization of 100 gm of each copolymer. There are several breaks before the complete neutralization of all phenolic hydroxyl groups (Michel et al., 2007). The first break in the plot was the smallest break and assumed that this corresponds to a stage in titration when an average one phenolic hydroxy group of each chain was neutralized. From the plot, the first and final breaks were recorded. The average degree of polymerization $(\overline{DP})$ and the number average molecular weight $(\overline{Mn})$ of all copolymers have been determined given below.

$$DP = \frac{\text{Total milli equivalents of base required for complete neutralization}}{\text{Milli equivalents of base required for smallest interval}}$$

$\overline{Mn} = \overline{DP}$ x Repeat unit weight

It is observed that the molecular weight of copolymers increase with increase in p-cresol content. This observation is in agreement with the trend observed by earlier workers (Michel et al., 2007; Shah et al., 2006).

Electronic Spectra

The UV-visible spectra of all the p-CAF copolymer samples in pure DMSO were recorded in the region 200–850 nm at a scanning rate of 100 nm/min and a chart speed of 5 cm/min. All of the four p-CAF copolymer samples displayed two characteristic broad bands at 325–335 and 260–285nm. Both of these bands seem to be merged with each other because of their very broad nature. These observed position for the absorption bands indicate the presence of a carbonyl (C=O; ketonic) group having a carbon–oxygen double which is in conjugation with -NH group. The former band (more intense) can be accounted for by $n \rightarrow \pi^*$ transition while the latter band (less intense) may be due to $n \rightarrow \pi^*$ transition (Gurnule et al., 2003a). The bathochromic shift (shift towards longer wave length) from the basic values of the >C=O group viz. 320 and 240 nm, respectively may be due to the combined effect of conjugation of >C=O and NH groups and phenolic hydroxyl group (auxochrome) and phenyl ring (Shah et al., 2006). It may be observed from the UV-visible spectra of p-CAF copolymers that the absorption intensity gradually increases in the order p-CAF-1 < p-CAF-2 < p-CAF-3 < p-CAF-4. The observed increasing order may be due to introduction of more chromophore (>C=O groups) and auxochrome (phenolic -OH) in the repeat unit structure of the copolymers.

Infrared Spectra

The IR spectral data are tabulated in Table 4. From the IR spectral studies, it has been revealed that all the four p-CAF copolymers give rise to nearly similar pattern of spectra. Very broad band appeared in the region 3220–3230 cm^{-1} may be assigned to the stretching vibration of phenolic -OH groups exhibiting intermolecular

Table 4. Molecular weight determination, viscometric data, and IR frequencies of the p-CAF copolymer resins.

Copolymer	Empirical formula of repeat unit	Empirical weight of repeat unit (g)	Average degree of polymerization $(\overline{DP})$	Average molecular weight $(\overline{Mn})$	Intrinsic viscosity [η] (dl/g)	Huggins' constant K_1	Kraemmer's constant K_2	$K_{1+}K_2$	Important IR frequencies: Wave number (cm^{-1})	Important IR frequencies: Assignments
p-CAF-1	$C_{15}H_{20}N_2O_3$	276	25.00	6902	0.44	0.253	0.251	0.504	3220-3230 b,st	-OH phenol
									3010-3011b,st	-CH stretching of aromatic ring.
p-CAF-2	$C_{23}H_{28}N_2O_4$	396	21.35	8455	0.98	0.258	0.255	0.513	2916-2917 b,st	Intra molecular hydrogen bonding
									1606-1607 sh, st	>C=O group of oxamide
									1502-1503	Aromatic ring
p-CAF-3	$C_{31}H_{36}N_2O_5$	516	22.96	11850	1.50	0.261	0.253	0.514	1480-1482 sh,st	>CH_2 bending (scissoring)
									1378-1379 sh,st	>CH_2 bending (twisting and wagging)
									1200-1220 sh,st	>CH_2 i.e. methyl bridge
									1286-1290 sh,st	>CH_2 (plane bending)
p-CAF-4	$C_{39}H_{44}N_2O_6$	638	24.54	15662	1.90	0.262	0.256	0.518	910-990 sh, w	1,2,3,5 substitution in benzene skeleton
									1097-1098 b,m	
									1148-1149	
									1220-1260	

Sh = sharp, b = broad, st = strong, m = medium, w = weak

hydrogen bonding between -OH and >C=O and NH group of amide (Michel et al., 2007). The bands obtained at 1200–1220 cm^{-1} suggest the presence of methylene ($-CH_2-$) bridges. A sharp strong peak at 1502–1503 cm^{-1} may be ascribed to aromatic skeletal ring breathing modes. The 1,2,3, and 5 tetra substitution of aromatic benzene ring can be recognized from sharp and medium/weak absorption bands appeared at 910–990, 1097–1098, 1148–1149, and 1220–1260 cm^{-1} respectively. The presence of C-H stretching of aromatic ring may be assigned as a sharp and strong band at 3010–3011 cm^{-1} which seems to be merged with very broad band of phenolic hydroxy group.

Nuclear Magnetic Resonance Spectra

The NMR spectra of all the four copolymers were scanned in DMSO-d_6. The spectral data are given in Table 5. From the spectra, it is revealed that all p-CAF copolymers gave rise to different pattern of NMR spectra, since each p-CAF copolymer possesses set of proton having different electronic environment. The chemical shift (δ ppm) observed is assigned on the basis of data available in literature (Patel et al., 2007). All the p-CAF copolymer samples show an intense weakly multiplate signals at 2.16–2.17 (δ) ppm may be attributed to methyl proton of Ar-CH_3 group. The medium singlet at 2.53–2.59 (δ) ppm may be due to the methylene proton of Ar-CH_2 Bridge. The singlet obtained in the region of 3.47(δ) ppm may be due to the methylene proton of Ar-CH_2-N moiety. The signals in the region 5.23(δ) ppm are attributed to protons and -NH Bridge. The weak multiplate signals (unsymmetrical pattern) in the region at 6.93–6.98 (δ) ppm may be due to terminal methylene group. The signals in the range at 7.54–7.92 (δ) ppm may be due to phenolic hydroxy protons. The much downfield chemical shift for phenolic -OH indicates clearly the intramolecular hydrogen bonding on -OH group (Singru et al., 2008). The signal at 0(δ) ppm is due to TMS (tetramethyl silane) the signal at 1.20–1.21 (δ) ppm may due to -CH_3-C ≡ moiety. The signal at 2.80 to 2.81(δ) ppm and 3.37 to 3.35 (δ) ppm may due to CH_2-Ná moiety. The signal at 4.60–4.20(δ) ppm may due to >CH-O group. The signal at 7.30–7.31 (δ) ppm may due to aromatic proton (Ar-H).

Table 5. The ^{1}H NMR spectral data of p-CAF copolymer resins.

Observed Chemical Shift (δ) ppm				Nature of proton assigned	Expected chemical shift (δ) ppm
p-CAF-I	p-CAF-II	p-CAF-III	p-CAF-IV		
1.20	1.21	1.21	1.21	-CH_3-C≡ moiety	1.00 to 1.50
2.16	2.16	2.17	2.16	Methyl proton Ar-CH_3 group	2.00 to 3.00
2.80	2.81	2.81	2.80	CH_2-N⟨ moiety	2.3 to 3.8
3.71	3.71	3.70	3.71		
2.59	2.56	2.58	2.58	Methylene proton of Ar-CH_2 moiety	2.00 to 3.00
3.37	3.36	3.35	3.35	Methylene proton of Ar-CH_2-N moiety	3.00 to 3.5
4.61	4.62	4.60	4.61	>CH-O	4.00 to 5.00

Table 5. *(Continued)*

Observed Chemical Shift (δ) ppm				Nature of proton assigned	Expected chemical shift (δ) ppm
p-CAF-I	p-CAF-II	p-CAF-III	p-CAF-IV		
5.25	5.22	5.23	5.24	Proton of -NH bridge	5.00 to 8.00
6.78	6.75	6.73	6.74	Terminal methylene	6.2 to 8.5
7.31	7.30	7.31	7.31	Aromatic proton (Ar-H)	7.00 to 8.50
8.80	8.72	8.84	8.7	Proton of phenolic – OH involved intramolecular hydrogen bonding	8.00 to 12.00

Scanning Electron Microscopy (SEM)

Figure 1 shows the SEM (Suzuki, 2002) micrographs of the pure p-CAF-I copolymer sample at 1500X and 3000X magnification (particle size is 20 μm). The morphology of resin exhibits growth of crystals from polymers solution corresponding to the most prominent organization in polymers on a large scale such as in size of few millimeters spherulites. The morphology of resin shows a fringed micelle model of the crystalline-amorphous structure. The extent of crystalline character depends on the acidic nature of the monomer. The micrograph of pure sample shows the presence of crystalline-amorphous layered morphology which is the characteristic of polymer. The monomers have crystalline structure but during condensation polymerization of some crystalline structure lost into amorphous morphology.

On the basis of the nature and reactive position of the monomers, elemental analysis UV-visible, IR, NMR spectral studies and taking into consideration the linear structure of other phenol -formaldehyde and the linear branched nature of urea-formaldehyde polymers, the most probable structure has been proposed for p-CAF-1 copolymer resin, has been shown in Figure 1. The morphology of the resin shows the transition between crystalline and amorphous nature, when compare to the other resin (Jadhao et al., 2005a), the p-CAF copolymer resin is more amorphous in nature, hence, higher metal ion-exchange capacity.

Ion-exchange Properties

Batch equilibrium technique developed by Gregor et al. and De Geiso et al. was used to study ion-exchange properties of p-CAF-1 copolymer resin. Seven metal ions Fe^{3+}, Cu^{2+}, Ni^{2+}, Co^{2+}, Zn^{2+}, Cd^{2+}, and Pb^{2+} in the form of aqueous metal nitrate solution were used. The ion-exchange study was carried out using three experimental variables: (a) Electrolyte and its ionic strength (b) uptake time, and (c) pH of the aqueous medium, Among these three variables, two were kept constant and only one was varied at a time to evaluate its effect on metal uptake capacity of the polymers similar to the earlier co-workers (Jadhao et al., 2005b; Shah et al., 2006; Singru et al., 2008). The details of experimental procedure are given below.

Effect of Electrolyte and Its Ionic Strength on Metal Uptake

We examined the influence of ClO_4^-, NO_3^-, Cl^-, and SO_4^{2-} at various concentrations on the equilibrium of metal-resin interaction. Figure 2–5 show that the amount of metal ions taken up by a given amount of copolymer depends on the nature and concentration of the electrolyte present in the solution. Generally as concentration of the electrolyte increases, the ionization decreases, number of ligands (negative ions of electrolyte) decrease in the solution which forms the complex with less number of metal ions and therefore more number of ions may available for adsorption. Hence on increasing concentration, uptake of metal ions may be increased, which is the normal trend. But this normal trend disturbed due to the formation of stable complex with more number of metal ions with electrolyte ligands, which decrease the number of metal ions available for adsorption, hence uptake decreases.

Electrolyte solution + metal ion solution + polymer → electrolyte ligand – metal ion chelates + Polymer – metal ion chelates

If electrolyte ligand – metal ion complex is weak than polymer metal ion chelates, the more metal ion can form complex with polymer hence uptake of metal ion is more. But if this complex is strong than polymer – metal ion chelates, more metal ions form strong complex with electrolyte ligand which make metal uptake capacity lower by polymer.

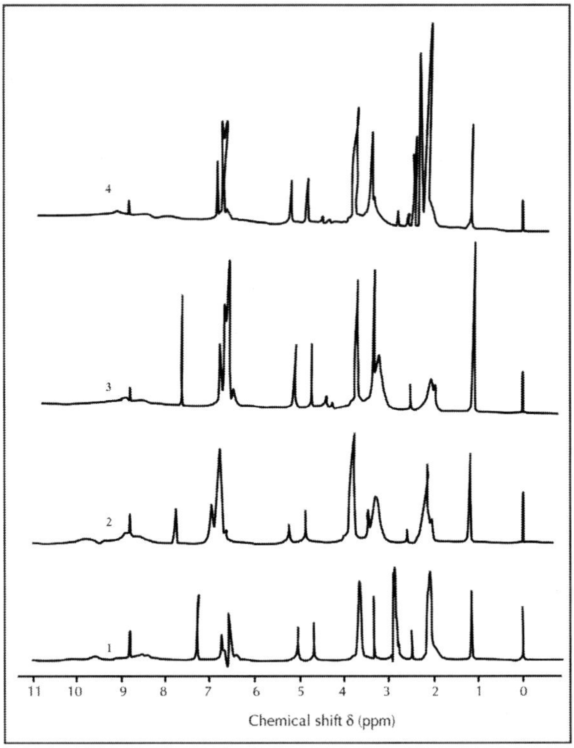

Figure 3. Nuclear magnetic resonance spectra of p-CAF copolymer resins (1) p-CAF-I, (2) p-CAF-II, (3) p-CAF-III, and (4) p-CAF-IV.

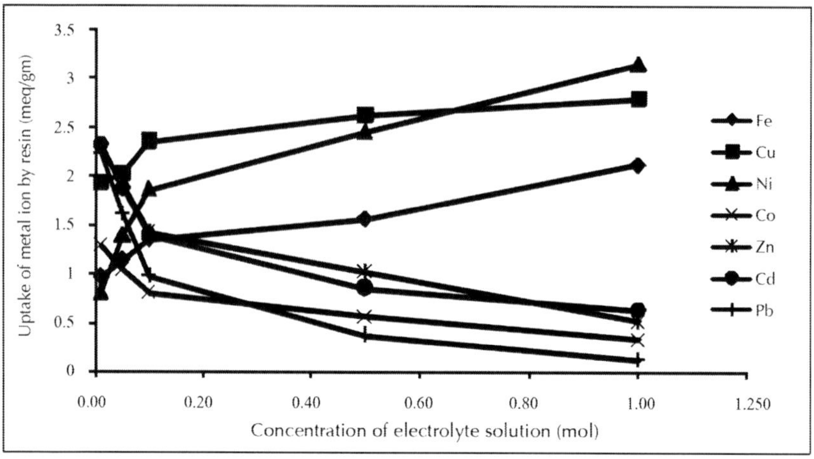

Figure 4. Uptake of several metal ions[a] by p-CAF-I copolymer resin at five different concentrations of electrolyte $NaNO_3$ solution.

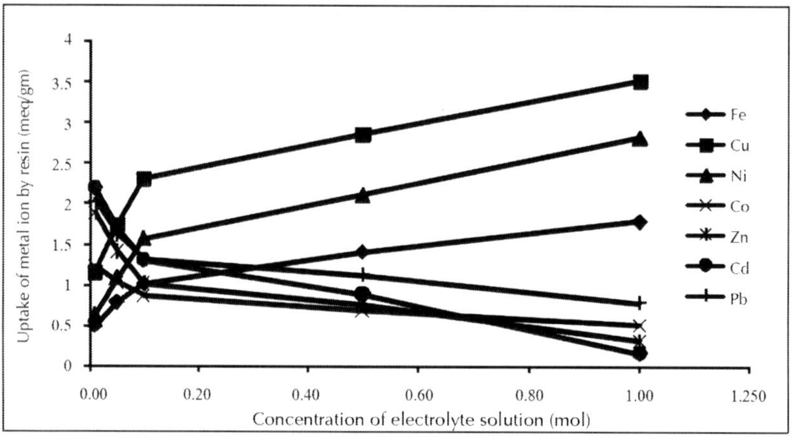

Figure 5. Uptake of several metal ions[a] by p-CAF-I copolymer resin at five different concentrations of electrolyte NaCl solution.

In the presence of perchlorate, chloride and nitrate ions, the uptake of Fe(III), Cu(II), and Ni(II) ions increase with increasing concentration of the electrolytes, whereas in the presence of sulfate ions the amount of the above mentioned ions taken up by the copolymer decreases with increasing concentration of the electrolyte (Jadhao et al., 2005b; Manavalan et al., 1983). Moreover, the uptake of Co (II), Zn (II), Cd (II), and Pb (II) ions increase with decreasing concentration of the chloride, nitrate, perchlorate and sulfate ions. This may be explained on the basis of the stability constants of the complexes with those metal ions (Jadhao et al., 2005b). The ratio of physical core structure of the resin is significant in the uptake of different metal ions by the resin polymer. The amount of metal ion uptake by the p-CAF-I copolymer resin is found to be higher when comparing to the other polymeric resins (Burkanudeen et al., 2002; Manavalan et al., 1983). The ratio of physical core structure of the resin is significant in the uptake of different metal ions by the resin polymer. The metal uptake order of p-CAF copolymer resins is found to be p-CAF-I < p-CAF-II < p-CAF-III < p-CAF-IV, for all metal ions. This sequence of metal uptake may be due to introduction of more and more aromatic cresol rings, -OH group in the repeating unit of copolymer resins from p-CAF-I to p-CAF-IV. Because increasing the number of cresol rings and -OH, the conjugation and delocalization increases which may increase the void space in the copolymer resin which therefore increase the metal uptake capacity due to increasing molar ratio.

Estimation of the Rate of Metal Ion Uptake as a Function of Time

The rate of metal adsorption was determined to find out the shortest period of time for which equilibrium could be carried to while operating as close to equilibrium conditions as possible. As shaking time increases the polymer gets more time for adsorption, hence uptake increases on the increasing in the time. Figure 6 show the dependence of the rate of metal ion uptake on the nature of the metal. The rate refers to the change in the concentration of the metal ions in the aqueous solution which is in contact with the given polymer. The result shows that the time taken for the uptake of the different mental ions at a given stage depends on the nature of the metal ion under given conditions. It is found that Fe (III) ions require about 3 hr for the establishment of the equilibrium, whereas Cu (II), Ni (II), Co (II), and Zn (II) ions required about 5 or 6 hr (Gurnule et al., 2003b; Jadhao et al., 2005b). Thus the rate of metal ions uptake follows the order Cu (II) > Ni (II) > Co (II) ≈ Zn (II) > Cd (II) > Pb (II) for all of the copolymers (Jadhao et al., 2005b; Manavalan et al., 1983). The rate of metal uptake may depend upon hydrated radii of metal ions. The rate of uptake for the post transition metal ions exhibit other trend for Cd (II), the rate of uptake is in the comparable that of Pb (II) because of difference in 'd' orbital.

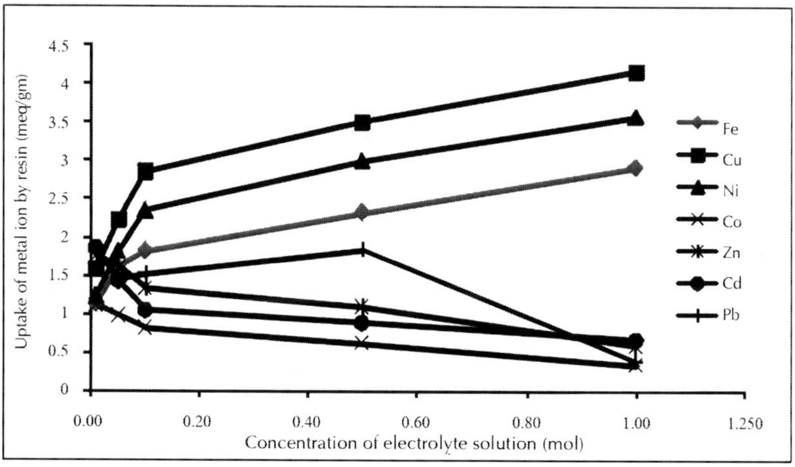

Figure 6. Uptake of several metal ions[a] by p-CAF-I copolymer resin at five different concentrations of electrolyte $NaClO_4$ solution.

Distribution Ratios of Metal Ions at Different pH

The effect of pH on the amount of metal ions distributed between two phases can be explained by Figures 7–9. The data on the distribution ratio as a function of pH indicate that the relative amount of metal ion taken up by the p-CAF copolymer increases with increasing pH of the medium (Burkanudeen et al., 2002; Gurnule et al., 2003b). The magnitude of increase, however, is different for different metal cations. The study was carried from pH 2.5 to 6.5 to prevent hydrolysis of metal ions at higher pH. For metal ion Fe^{3+} the highest working pH is 3, where distribution ratio is medium, since Fe^{3+} forms octahedral complex with electrolyte ligand, showing crowding effect (sterric hindrance), which may lower the distribution ratio of Fe^{3+} ions. The value of distribution ratio at particular pH thus depends upon the nature and stability of chelates with particular metal ion. The data of distribution ratio show a random trend in certain cases (Manavalan et al., 1983). This may be due to the amphoteric nature of the p-CAF resin. The following structure has been proposed for the polychelate (Figure10).

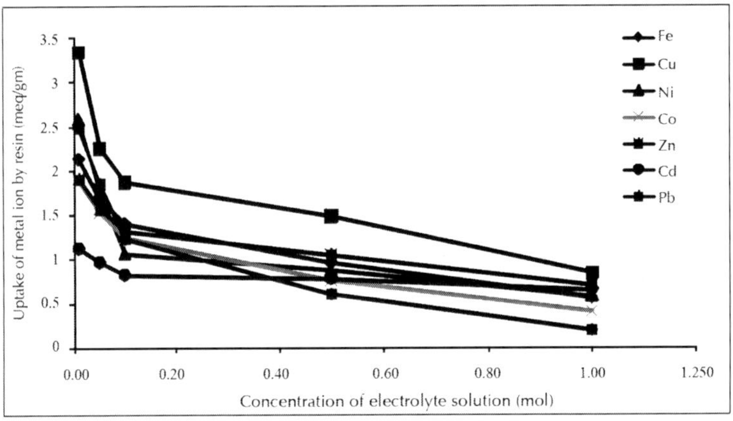

Figure 7. Uptake of several metal ions[a] by p-CAF-I copolymer resin at five different concentrations of electrolyte Na_2SO_4 solution.

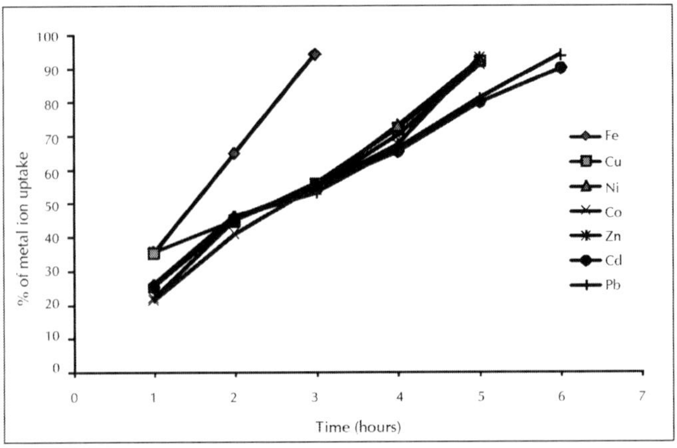

Figure 8. Comparison of the rate of metal ion[a] (m) uptake[b] by p-CAF-I copolymer resin.

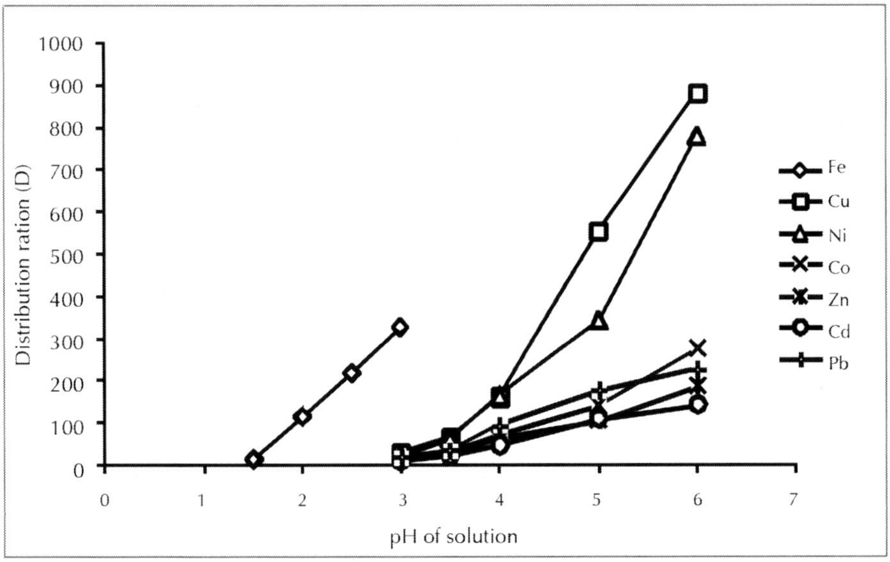

Figure 9. Comparison of the rate of metal ion[a] (m) uptake[b] at different pH.

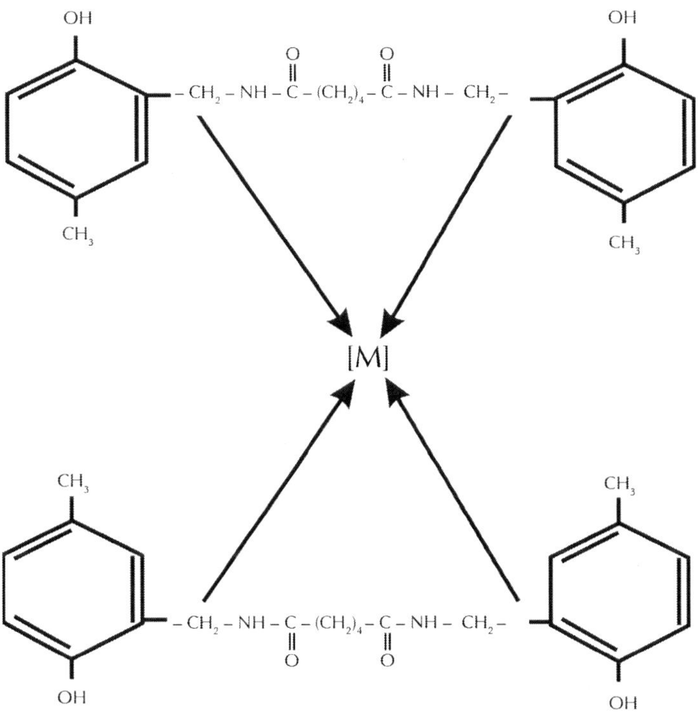

Figure 10. Proposed structure of polymeric chelate.

From the result it reveals that with decrease in atomic number, the ion uptake capacity is increased. In case of Cd (II) and Pb (II) purely electrostatic factors are responsible. The ion uptake capacity of Cd (II) is lower owing to the large size of its hydrated ion than that of Cu (II). The sterric influence of the methyl group and hydroxyl group in p-CAF resin is probably responsible for their observed low binding capacities for various metal ions. The higher value of distribution ratio for Cu (II) and Ni (II) at pH 4–6.0 may be due to the formation of more stable complex with chelating ligands. Therefore the polymer under study has more selectivity of Cu^{2+} and Ni^{2+} ions at pH 4.0–6.0 then other ions which form rather weak complex. While at pH 3 the copolymer has more selectivity of Fe^{3+} ions. The order of distribution ratio of metal ions measured in pH range 2.5–6.5 is found to be Fe (III) > Cu (II) > Ni (II) > Zn (II) > Co (II) > Pb (II) > Cd (II). Thus the results of such type of study are helpful in selecting the optimum pH for a selective uptake of a particular metal cation from a mixture of different metal ions (Singru et al., 2008). For example, the result suggests the optimum pH 6.0, for the separation of Co (II) and Ni (II) with distribution ratio 'D' at 415.4 and 854.4 respectively using the p-CAF copolymer resin as ion-exchanger. Similarly, for the separation of Cu (II) and Fe (III) the optimum pH is 3, at which the distribution ratio 'D' for Cu (II) is 66.1 and that for Fe (III) are 341.5. The lowering in the distribution of Fe (III) was found to be small and, hence, efficient separation could be achieved (Burkanudeen et al., 2002; Gurnule et al., 2003b).

The strength of ion-exchange capacities of various resins can be studied by comparing their ion-exchange capacities. The ion-exchange capacity (IEC) is a fundamental and important quantity for the characterization of any ion-exchange material. It is defined as the amount of ion that undergoes exchange in a definite amount of material, under specified experimental conditions. The ion-exchange capacity of p-CAF copolymer has been calculated, which was found to be 4.3 mmol/g which indicates that p-CAF copolymer resin is better ion-exchanger than commercial phenolic and some polystyrene commercial ion-exchangers.

The observation obtained indicates that, time required for the rate of metal ion uptake depends on the nature of the metal ions and may be depend on the ionic size. Thus the rate of metal ion uptake follows the order–

Metal ion	$Fe^{3+} > Cu^{2+} \approx Ni^{2+} > Co^{2+} \approx Zn^{2+} > Cd^{2+} \approx Pb^{2+}$						
Ionic size	0.55	0.57	0.69	0.90	0.90	1.10	1.19

For the strongly acidic cation exchange resin such as cross linked polystyrene sulphonic acid resins, the ion-exchange capacity is virtually independent of the pH of the solutions. For weak acid cation exchangers, such as those containing carboxylate group, ionizations, occurs only in alkaline solution. Similarly weakly basic cation exchanger does not work above pH 9.

CONCLUSION

A copolymer p-CAF-I based on the condensation reaction of p-cresol and adipamide with formaldehyde in the presence of acid catalyst was prepared. The p-CAF is a selective chelating ion-exchange copolymer resin for certain metals. The copolymer resin showed a higher selectivity for Fe^{3+}, Cu^{2+}, and Ni^{2+} ions than for Co^{2+}, Zn^{2+}, Cd^{2+}, and Pb^{2+} ions. The uptake of some metal ions by the resin was carried out by the batch equilibrium technique. The uptake capacities of metal ions by the copolymer resin were pH dependent. From the results of distribution coefficients, it can be observed that Cu (II) has higher value of distribution ratio. Due to considerable difference in the uptake capacities at different pH and media of electrolyte, the rate of metal ion uptake and distribution ratios at equilibrium, it is possible to separate particular metal ions from their admixture by this technique.

KEYWORDS

- **Absorption**
- **Degree of polymerization**
- **Ion-exchangers**
- **Metal-polymer complexes**
- **Resins**
- **Selectivity**

ACKNOWLEDGMENT

The authors are pleased to express their gratitude to the Director, Laxminarayan Institute of Technology, Rashtrasant Tukadoji Maharaj Nagpur University, Nagpur, India, for providing the necessary laboratory facilities.

Chapter 3

High Performance Shear Stable Viscosity Modifiers

B. Khemchandani and H. S. Verma

INTRODUCTION

A set of four Viscosity Modifier (VM) or Viscosity Index Improvers (VII) and their blends in American Petroleum Institute (API) group I to group IV base oils have been studied for key performance properties that is thickening tendencies, viscosity index improvement, shear stability, and shear stability indices. All the four base oils have been characterized by Infra Red spectroscopy for their chemical composition. Gel Permeation Chromatography (GPC) has been used for characterizing the molecular weight distribution of VMs.

Model equations using multiple regression technique for correlating VMs performance with some of the characteristics data of VMs as obtained from GPC and chemical composition of base oils have been obtained. Although statistical techniques are being used in many areas, this is relatively an unexplored area for predicting the performance of VMs where encouraging results with high accuracy have been obtained.

Oil soluble polymeric additive, known as VM or VII, represent an important category of base oil additives, which increases the viscosity of the formulated blend (Singh et al., 1987; Warren et al., 2005). The use of appropriate polymeric additive allows a lubricating oil to meet certain high temperature viscosity target while maintaining its low temperature flow property. The VII are usually high molecular weight polymer, broadly classified into two general categories: Olefin Based Copolymer (OCP) including Polystyrene and Hydrogenated Diene and Ester based which includes Polymethaacrylate (PMA). When dissolved in base oil, these high molecular weight polymers increase the viscosity index of the base oils.

Over the last three decades, many researchers have studied the viscometric properties of the multigrade industrial and automotive oils over a wide temperature range using various combinations of VII in different base stocks (Abou El Naga et al., 1998; Alexander et al., 1989; Coutinho et al., 1993). One of the major problems with many of the VII is their tendency to undergo permanent reduction of viscosity as a result of mechanical shearing which occur in most of the mechanical system (Bartz, 1999; Hassanean et al., 1994; Jain et al., 2000).

The role of VII or VMs in industrial lubricants is a complex one. While they are known to alter both the viscosity and viscosity index of lubricant base stocks at low shear, their contribution under the high pressure and shear rate is more difficult to assess (Wardle et al., 1990).

The API has defined the base oils in five categories on the basis of viscosity index, sulfur content and percentage saturates. The details are provided in Table 1. A number of studies in past decade have been performed on various polymers in conventional base stocks (API group I), however with the introduction of API group II and API group III oils, a need was felt to analyze the performance properties of commercially available VMs in these new classes of base oils.

Table 1. API base oil category.

Base oil Category	Sulphur (%)	Saturates(%)	Viscosity Index
Group I	> 0.03% AND / OR	< 90	80 to 120
Group II	< = 0.03 AND	> = 90	90 to 120
Group III	< = 0.03 AND	> = 90	> = 120
Group IV	ALL POLYALPHAOLEFINS (PAO's)		
Group V	All Other Base Stocks Not Included In GROUP I,II,II,IV		

In the present work, authors have characterized commercially available four API group I to group IV base stocks and four VMs. Blends of four polymers in the entire base stocks were studied for a number of performance parameters including shear stability. A multiple regression analysis has also been done to correlate the viscosity index increase, shear stability and shear stability index with the chemical composition of base stocks and polymers.

EXPERIMENTAL

Base Oil Characterization

Four 150 Neutral Base Oils of different degree and type of refining (API group I to IV) have been used for formulating the desired blend. These base oils have been characterized for paraffinic, aromatic and napthenic contents by Infra Red spectroscopy technique and sulfur content by X-ray fluorescence technique. Brief characteristics of these base oils have been summarized in Table 2.

Table 2. Characteristics properties of base oils.

Properties	Base Oil 1 (Group I)	Base Oil 2 (Group II)	Base Oil 3 (Group III)	Base Oil 4 (Group IV)
Appearance	Clear	Clear	Clear	Clear
Colour	<1.0	<0.5	<0.5	<0.5
Density @ 15 ^{0}C, g/cm^3	0.8758	0.8599	0.8492	0.8270
Kinematic Viscosity @ 40 ^{0}C, cSt	29.03	32.7	34.82	30.60
Kinematic Viscosity @ 100 ^{0}C, cSt	5.08	5.51	6.15	5.83
Viscosity Index	102	105	126	137
Pour Point, ^{0}C	(-) 6	(-) 21	(-) 27	(-) 51
'S' by XRF, ppm	6000	10	10	<10
Hydrocarbon Analysis				
Ca	7.08	2.98	0.34	0.34
Cp	78.24	73.69	78.81	95.75
Cn	14.68	23.33	20.85	3.91

Characterization of Additive

Four commercial VMs or VII of PMA type, available in the Indian Market have been selected and studied by Infra Red spectroscopy for their chemical type. The GPC has been used to determine the molecular weight distribution of this VMs. The data obtained by GPC has been provided in Table 3.

Table 3. Characterization of additive.

Properties	VM 1	VM 2	VM 3	VM 4
Appearance	Viscous, Solubilised liquid	Viscous, Solubilised liquid	Viscous, Solubilised liquid	Viscous, Solubilised liquid
M_w X 10^3	120	104	46	52
M_N X 10^3	56	50	22	27
Poydispersity	2.14	2.08	2.09	1.92
Polymer Content	44.38	66.39	72.26	70.05

Viscosity Modifier Blends in Base Oils

The 8% of each polymer has been dissolved in the entire API Group I to IV base oils and thus obtained 16 combinations. The blending was performed by mechanical stirring at a temperature of 70–80°C, for 30 min. Kinematic viscosity as per ASTM D445 test procedure on all the 16 blends using Cannon Routine Viscometer (Model CAV 200, Cannon Fully Automated Digital Kinematic Viscometer) have been carried out at 40°C +/– 0.01 and 100°C +/– 0.01. This test method is used to measure the time for a volume of liquid to flow under the gravity through a calibrated glass capillary viscometer. The viscosity index of these blends has been calculated using ASTM D2270 test procedure.

Shear Stability Conducted on Blends

This test method was performed as per ASTM D6278 test procedure that covers the evaluation of the shear (mechanical stress) stability of polymer containing fluids. The polymer containing fluid is passed through a fine nozzle at a shear rate that causes the less shear stable molecule to degrade. In this method, 150 ml of the oil sample is filled in the fluid reservoir and the sample is passed through Bosch injector nozzle at an ambient temperature at a pressure of 175 bar. The resultant degradation reduces the kinematics viscosity of the fluids under test. The test method measures the percent viscosity loss at 100°C of polymer containing fluids.

RESULTS AND DISCUSSION

Thickening Tendency

Thickening tendency is one of the important parameter of any polymer. A polymer is an effective viscosity index improver when the numerical value of it is more than 1. Thickening tendency is the ratio of specific viscosity at 100°C to specific viscosity at 40°C.

Specific viscosity is calculated as per given below Equation (1).

Specific viscosity, $(Vsp) = \frac{(V - Vo)}{Vo}$ (1)

where,

Vo = Viscosity of the base oil and

V = Viscosity of the polymer containing blend.

Specific viscosities at 100°C and 40°C has been obtained using Equation (1) and the values are depicted in Figure 1 and Figure 2 respectively. The thickening tendencies as calculated from the experimental values of specific viscosities for all the polymers in different base oils are given in Figure 3.

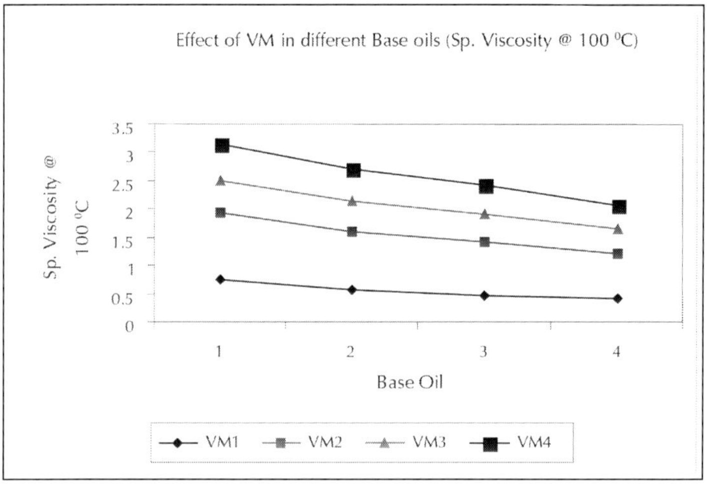

Figure 1. Specific viscosities at 100°C.

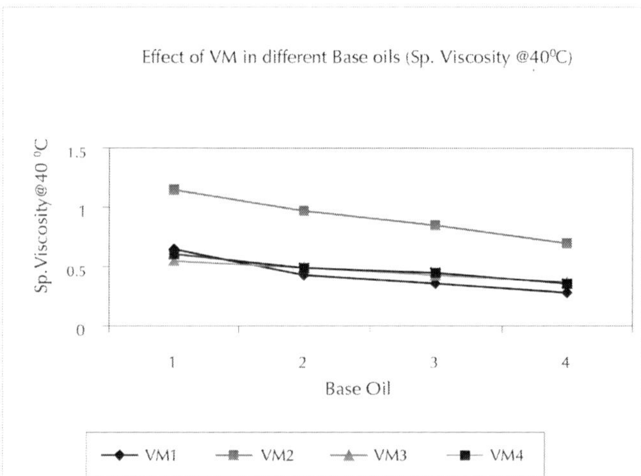

Figure 2. Specific viscosities at 40°C.

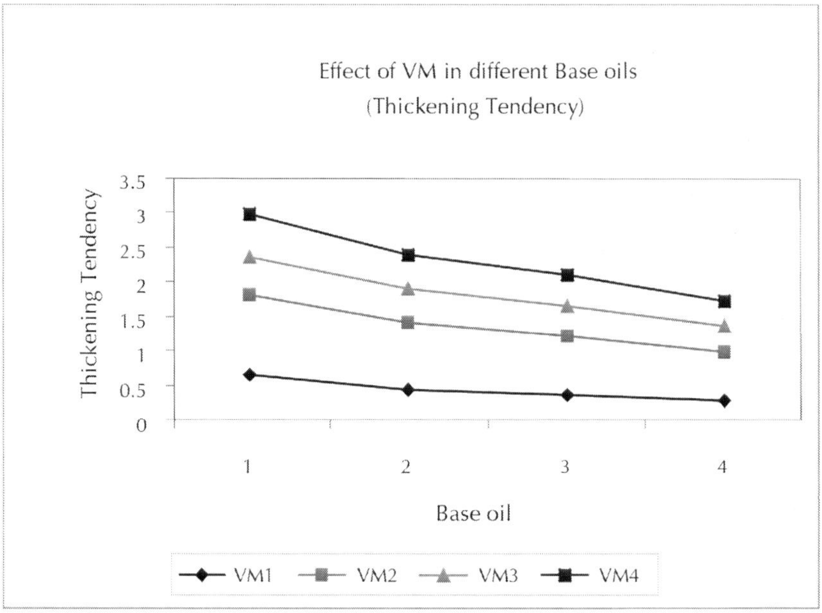

Figure 3. Thickening Tendencies of viscosity modifiers in API group I to IV base oils.

It is evident from the Figure 3 that the thickening tendencies for all four polymers in group I to group IV base oils is different and it is steadily decreasing from API group I oil to group IV oils. The specific viscosities at 100°C is higher in case of VM 3 and VM 4, however, these polymers have lower specific viscosities at 40°C and therefore have higher thickening tendencies.

Viscosity Index Increase

The main purpose of adding VMs in base stocks is to increase the viscosity index of the resultant industrial product and therefore it is also one of the essential and important parameter of the VMs. Viscosity index increase has been observed for different VMs as well as it have also been found that it also varies from base oil to base oil.

Shear Stability

Shear stability is one of the parameter that is used for assessing the performance of the polymer under standard test condition.

The shear stability of the blend, containing polymer is calculated using Equation (2).

$$\text{Shear stability} = \left[\left(\frac{V_2}{V_1}\right)\times 100\right] \quad (2)$$

where,

V_1=Kinematic viscosity of the viscosity index improver containing oil before shearing

V_2=Kinematic viscosity of the viscosity index improver containing oil after shearing

Shear Stability Index

Shear stability index is an indicator of stability of the industrial oil under severe stress condition and it is assessed as per standard test procedure ASTM D6278. Lower the shear stability index better is the formulated Industrial Lubricant and It is calculated using the given below Equation (3).

$$\text{Shear stability index} = \left[\frac{(V_1 - V_2)}{(V_1 - V_O)} \times 100\right] \quad (3)$$

where,

V_O = Base oil viscosity

V_1 = Kinematic viscosity of the viscosity index improver containing oil before shearing

V_2 = Kinematic viscosity of the viscosity index improver containing oil after shearing

Regression Analysis

For prediction of viscosity index increase, shear stability and shear stability index, a multiple regression analysis has been carried out using backward elimination technique. The regression equations for estimating viscosity index increase, shear stability and shear stability index are given by Equation (4), (5), and (6) respectively. A very high value of correlation coefficients of 0.978, 0.905, and 0.896 have been obtained indicating the model equations can be used for estimating the viscosity index increase, shear stability and shear stability index respectively with great accuracy. Further, a fairly low standard error of estimation of 1.7, 0.79, and 1.7 has been obtained for viscosity index increase, shear stability and shear stability index respectively, indicating that the theoretical results obtained are well within the repeatability.

$$\text{Viscosity index increase} = 122.5476 + 0.34713 \times Ca - 0.388021 \times Cp - 0.310172 \times Cn + 1.517745 \times Mw - 0.421034 \times PD + 0.567201 \times PC \quad (4)$$

$$\text{Shear stability} = 120.6588 + 0.03137 \times Ca - 0.13497 \times Cp - 0.31535 \times Cn - 1.66456 \times Mw + 0.01186 \times PD - 1.05694 \times PC \quad (5)$$

$$\text{Shear stability index} = -34.6323 - 0.071858 \times Ca + 0.268490 \times Cp + 0.434058\, Cn + 1.504772 \times Mw - 0.060587 \times PD + 0.705001 \times PC \quad (6)$$

where,

Ca = Aromatic carbon % of base oil,

Cp = Paraffinic carbon % of base oil,

Cn = Napthenic carbon % of base oil,

Mw = Weight average molecular weight in k Dalton of the polymer,

PD = Polymer dispersivity and

PC = Polymer content in the commercial polymer

Predicted and experimental values of viscosity index increase, shear stability and shear stability index have been shown in Figure 4, Figure 5, and Figure 6, respectively.

As it is clearly evident from Figure 4, Figure 5, and Figure 6 that the performance parameters such as viscosity index increase, shear stability and shear stability index of the VMs depends on chemical composition of base oils and polymers and the above equations can be used for fair theoretical estimation of these characteristics.

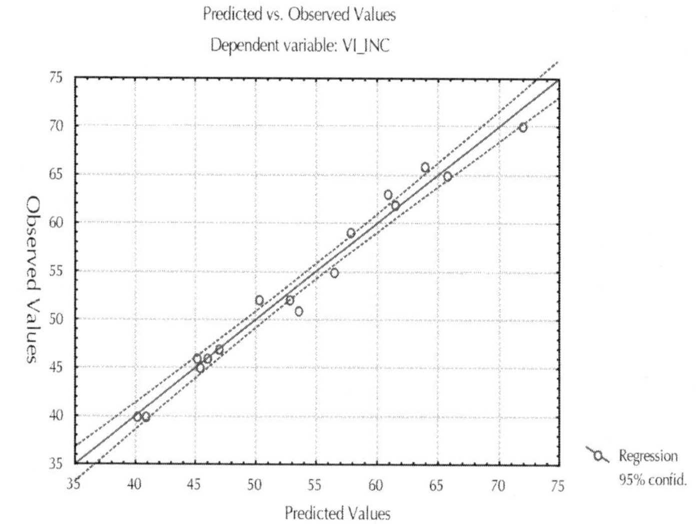

Figure 4. Predicted vs. Observed values for Viscosity Indices Increase.

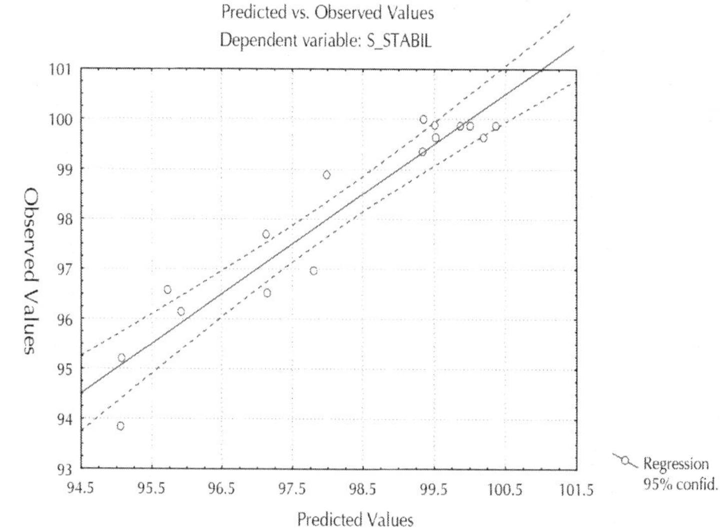

Figure 5. Predicted vs. Observed values for Shear Stability.

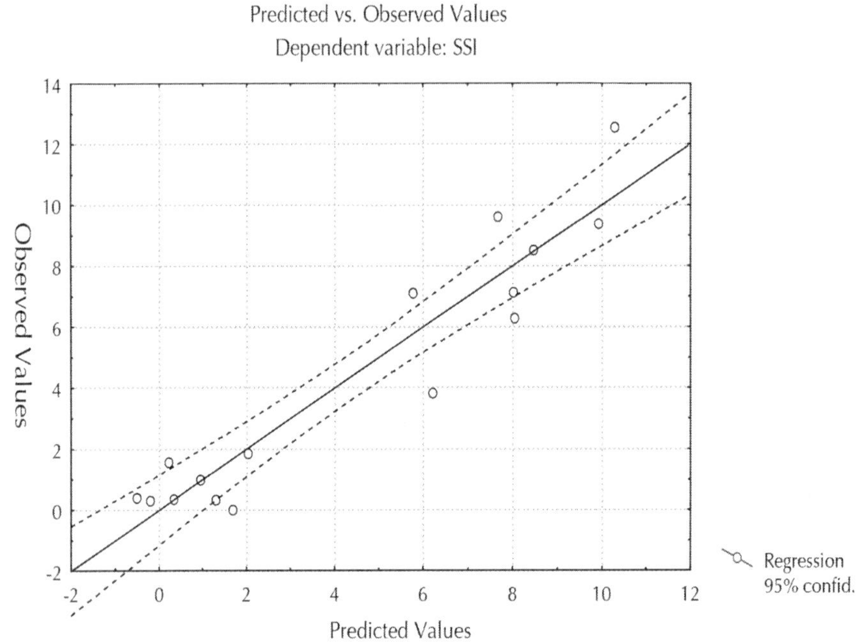

Figure 6. Predicted vs. Observed values for Shear Stability Index.

From Equation (4) and (6), the viscosity index increase and shear stability index for a specific VM in a particular base oil is higher if the molecular weight and concentration of the polymer is higher, however from Equation (5) the shear stability of the VM is lower if the molecular weight and concentration of VM is higher.

Shear stability index is still widely used for assessing the stability of polymer and therefore its estimation is must to ensure that the polymer or VM is suitable for use in Industrial Lubricants. The above model equation will allow a quick estimation of the shear stability index of the VM in any of the API group I to group IV base oils using composition of base oil and molecular weight distribution of the VM.

CONCLUSION

For each VMs, there is a continuous decrease in the specific viscosity and thickening tendency from API group I to group IV base oils indicating that the efficiency of the VM is highest in API group I and lowest in group IV because of presence of aromatics that is >5%.All VMs show varying viscosity index increase, shear stability and shear stability index in different base oils and therefore these parameters not only depends on the characteristics of VMs but also depends on base oil composition (Paraffin, napthenes and aromatics).

Model equations for correlating the performance of VMs with high correlation coefficients (0.978 – 0.896) for estimation of key performance parameters of VMs that is viscosity index increase, shear stability and shear stability index have been obtained.

These equations can suitably be used for screening the key performance properties of VMs especially shear stability and shear stability indices allowing the Industrial Lubricant formulator to choose a right VM that remains reasonably stable in modern industrial machines operating under high mechanical stress.

KEYWORDS

- **Base oils**
- **Correlation**
- **Multiple Regressions**
- **Shear stability**
- **Shear Stability Index**
- **Viscosity Index Improvers**
- **Viscosity Modifiers**

Chapter 4

Eco-friendly Synthesis of Phthalonitrile Polymers

Palaniappan Selvakumar, Muthusamy Sarojadevi, and Pudupadi Sundararajan

INTRODUCTION

A novel, efficient methodology for the synthesis of phthalonitrile derivatives was investigated, using ionic liquid (IL) and microwave (MW) media as well as both simultaneously. Phthalonitrile monomers containing imide linkages were prepared from the reaction between aromatic dianhydrides, 3,3',4,4'-benzophenonetetracarboxylic dianhydride (BPTA), pyromellitic dianhydride (PMDA), 4,4'-(hexafluroisopropylidene) diphthalic anhydride (6FDA), 3,3',4,4'-biphenyltetracarboxylic dianhydride (BTDA) and the end-capping agent 4-(3-aminophenoxy) phthalonitrile through the imidization reaction. The use of the IL 1-butyl-3-methyl imidazoliumchloride as a solvent significantly increased the rate and yield of the reaction. The use of MW irradiation and reaction parameters significantly shortened the reaction time while enhancing the purity. The polymerization of the prepared phthalonitrile monomers was carried out with 3 wt % of aromatic diamine and 4,4'-oxydianiline (ODA) curing agent under MW irradiation. It is shown that condensation was successfully carried out using the recyclable IL medium under MW irradiation.

Most of the chemical reactions involve organic solvents which are volatile organic compounds (VOCs), toxic to various extent and non-recyclable. Developing alternative eco-friendly green solvents for chemical reactions and polymerization processes is an emerging area. Recently ILs have been introduced as a new class of recyclable solvents. These solvents are mostly liquid at room temperature, and contain entirely of ionic species. They have many interesting properties such as nonvolatility, good conductivity, good thermal stability, nonflammability and can dissolve most organic and inorganic materials. These attributes make them appealing candidates for chemical reactions (Earle et al., 2000; Holbrey et al., 1999; Hou et al., 2007; Kubisa, 2004). In recent years, the application of ILs has been explored as polymerization media as well as catalysts (Kumar et al., 2007). Substituted imidazolium group ILs have been widely used for this type of applications since these cations are comparatively low melting salts due to its low charge density (Yeganeh et al., 2004). In addition, ILs readily absorb MW energy and can be heated rapidly due to its high polarity and electrical conductivity.

Other important aspects are the reaction time and heating source. Generally chemical reactions need even and efficient heating. The conventional heating techniques are often slow and time consuming, and sometimes could lead to overheating and decomposition of the substrate and the product. It is well known that MW enhances reaction rates, increases selectivity and yields. The MW irradiation technique provides an efficient and fast synthetic route by heating molecules directly through the interaction between the MWs and molecular dipole moments of the reactants. There are two ways in which MWs can heat the reactant molecules, dipolar polarization (dipole relaxation) and ionic conduction (Giribabu et al., 2007; Kappe, 2004). Weataway and Perreux et al. (Perreux et al., 2001; Westaway et al., 1995) suggested the possibility of "non-thermal MW effects"/athermal effects with specific molecules during the reaction. Non-thermal effect is defined as the direct interaction between the electric field and specific molecules of the reaction medium which leads to orientation effects of dipolar molecules. Therefore, the reactant molecules could be treated both ways through MWs under IL medium. Different types of polymerization reactions such as radical, ring opening, step growth and condensation under MW conditions have already been reported (Guzman-Lucero et al., 2006; Lidstrom et al., 2001; Perreux et al., 2001; Westaway et al., 1995). The presence of ions/polar molecules is necessary for reactant molecules to be heated, under the MW irradiation. The ILs satisfies these requirements and should be considered as predominantly feasible for energy dissipation with MWs (Hoffmann et al., 2003).

The syntheses of polyimides and most of the other high performance polymers invariably involve high temperatures for extended period of time. The MW is an efficient alternate heating source for high temperature polymeric reactions. Phthalonitrile-based resins belong to a category of high temperature polymers with a variety of potential applications such as composite matrix materials (Brunel et al., 2008; Wiesbrock et al., 2004), adhesives (Dominguez et al., 2004) and electrical conductors (Sastri et al., 1997). These polymers exhibit excellent properties such as high thermal, oxidative and dimensional stability, mechanical properties, and low moisture absorption. Keller et al. have synthesized phthalonitrile polymers, having many different characteristics, containing ether (Keller, 1987, 1988), sulfone (Keller et al., 2005; Laskoski et al., 2005), and imide (Keller et al., 1984) linkages. Among these resins, preparation of imide linkage containing phthalonitrile resins involves very high temperatures and several hours of reaction time. The polymerization reaction with a small amount of (3–5 wt%) of aromatic diamine curing agent takes place for several hours (30–48 hr), in these conventional methods (Keller et al., 1985; Laskoski et al., 2007). In addition, it is also difficult to obtain a fully cured resin, because of the volatility of curing agent during the prolonged and uneven heating of monomers. Hence, it is necessary to explore approaches to increase the processability, and decrease the polymerization reaction time with an efficient alternative heating.

In this chapter, we describe the synthesis of imide linkage containing phthalonitrile monomers with three different reaction conditions and compare the results with the conventional methods.

EXPERIMENTAL

Materials

The 3,3',4,4'-benzophenonetetracarboxylic dianhydride (BPTA), PMDA, 4,4'-(hexafluroisopropylidene) diphthalic anhydride (6HIDA), 3,3',4,4'-biphenyltetracarboxylic dianhydride (BTDA), 1-butyl-3-methylimadazolium chloride and isoquinoline were purchased from Aldrich Chemicals, Canada. m-cresol was distilled under nitrogen before use. The 4-(3-aminophenoxy) phthalonitrile was prepared by the reported procedure (Keller, 1993).

Measurements

Fourier Transformation Infrared (FT-IR) spectra were recorded using a Perkin-Elmer RX-1 spectrometer with KBr pellet from 4,000 to 400 cm^{-1}. The ^{1}H NMR and ^{13}C NMR spectra were acquired at 300 MHz on a Bruker-300 spectrometer with 1% tetramethylsilane (TMS) as an internal standard. The DSC analysis was carried out with a Q_{10} series TA instruments differential scanning calorimeter using 5–7 mg of the sample crimped in aluminium pans at a heating rate of 10°C/min and under nitrogen atmosphere with a flow rate of 40 ml/min. The MW reactions were carried out in a Milestone Inc., laboratory MW system with a frequency of 2,450 MHz and controllable power system (max 1,000 W). A 50 ml (diameter 5 cm) Teflon reaction vessel was used. The temperature and time of the reaction were controlled by pre-programmed "Easywave" software system.

SYNTHESIS AND CHARACTERIZATION OF MONOMERS

Synthesis of Phthalonitrile Terminated Imide Monomer in Ionic Liquid (IL) Medium

To a 100 ml round-bottom flask was charged 3,3′,4,4′-benzophenonetetracarboxylic dianhydride (0.2 g; 000625 mol), phthalonitrile end-capping agent 4-(3-aminophenoxy) phthalonitrile (0.27 g; 0.00125 mol), 2 drops of isoquinoline catalyst and 5 g of IL 1-butyl-3-methylimadazolium chloride as a solvent. The reaction flask was purged with nitrogen and the solution was stirred for 3 hr at room temperature. The reaction mixture was heated slowly to reflux with stirring at 180°C for about 3 hr. The mixture was then cooled to room temperature and precipitated into excess water. The monomer was separated by filtration, washed with water several times to remove the IL and catalyst and dried overnight in a vacuum oven at 90°C. The FT-IR (KBr; cm^{-1}) 2232 (C≡N stretching), 1780 and 1716 (C=O asym and symm stretching), 1385 (C-N stretching), 1250 (C-O-C stretching).

Synthesis of Phthalonitrile Terminated Imide Monomer under Microwave (MW) Irradiation

A mixture of PMDA (0.146 g; 0.000625 mol) and phthalonitrile end-capping agent 4-(3-aminophenoxy) phthalonitrile (0.27 g; 0.00125 mol) was dissolved in 10 ml of m-cresol and two drops of isoquinoline were then added as a catalyst. After dissolving well the reaction mixture was carefully transferred to the MW reactor vial (bomb). The vials were moved in an automated fashion by a gripper attached to the platform. The reaction time and temperature were optimized by the trail reaction carried out with three different increasing durations of time. Then the reaction was performed at 180°C for about 15 min (180°C→15 min). The TLC analysis was performed to monitor the reaction. The mixture was cooled to room temperature and precipitated into excess ethanol. The monomer was separated by filtration, washed with ethanol several times to remove m-cresol and catalyst and dried overnight in a vacuum oven at 90°C. The FT-IR (KBr; cm^{-1}) 2232 (C≡N stretching), 1780 and 1716 (C=O asym and symm stretching), 1385 (C-N stretching), 1250 (C-O-C stretching)

Synthesis of Phthalonitrile Terminated Imide Monomers in Ionic Liquid (IL) under Microwave (MW) Irradiation

After dissolving the reaction mixture as above in IL, it was transferred to the MW reactor vial (bomb). The reaction was then completed as above.

Spectroscopic Data of Monomers

Monomer I

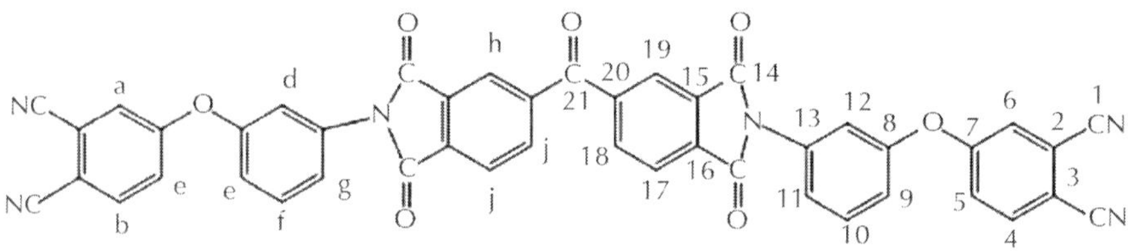

The FT-IR (KBr; cm^{-1}) 2230 (C≡N stretching), 1772 and 1712 (C=O asymm and sym stretching), 1365 (C-N stretching), 1251 (C-O-C stretching); ^{1}H NMR (300MHz, DMSO-d_6) δ (ppm) = 7.79, 7.88 (d, 2H, H_a), 8.14–8.22 (m, 6H, $H_{b, h, i}$), 7.49–7.52 (d d, 2H, H_c), 7.30–7.36 (m, 4H, $H_{d, g}$), 7.40–7.44 (d, 2H, H_e), 7.76–7.70 (t, 2H, H_f), 8.25–8.28 (d, 2H, H_j). The ^{13}C NMR (300MHz, DMSO-d_6) δ (ppm)= C_1- 117.4, C_2- 117, C_3- 113.2, C_4- 134.2, C_5- 123.7, C_6- 121.6, C_7- 161.1, C_8- 154.3, C_9- 119.3, C_{10}- 129.7, C_{11}- 120.9, C_{12}- 109.2, C_{13}- 132.3, C_{14}- 166.9, C_{15}- 132.5, C_{16}- 135.8, C_{17}-122.3, C_{18}-133.9, C_{19}-125.6, C_{20}-142.3, C_{21}-194.3.

Monomer II

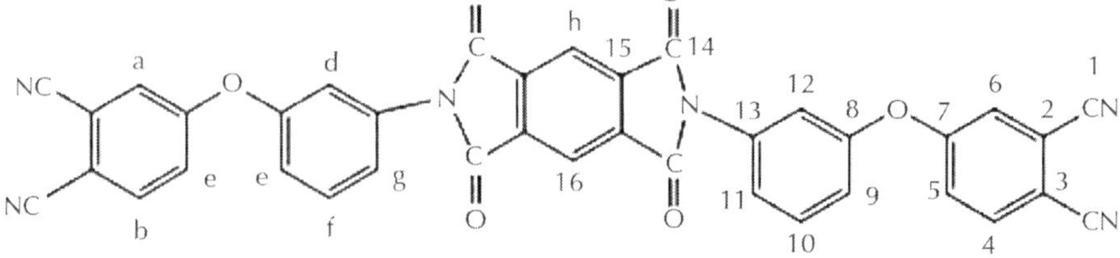

The FT-IR (KBr; cm^{-1}) 2232 (C≡N stretching), 1780 and 1716 (C=O asym and symm stretching), 1385 (C-N stretching), 1250 (C-O-C stretching); ^{1}H NMR (300MHz, DMSO-d_6) δ (ppm)= 7.54–7.46 (m, 4H, $H_{a, f}$), 8.18 (d, 2H, H_b), 7.92 (d, 2H, H_c), 7.41–7.32 (m, 4H, $H_{d, g, e}$), 8.41 (s, 2H, H_h). The ^{13}C NMR (300MHz, DMSO-d_6) δ (ppm) = C_1- 118.5, C_2- 117.3, C_3- 116.2, C_4- 133.6, C_5- 124.8, C_6- 123.2, C_7- 160.7, C_8- 154.6, C_9- 119.1, C_{10}- 131.7, C_{11}- 120.4, C_{12}- 115.8, C_{13}- 136.9, C_{14}- 165.5, C_{15}-137.5, C_{16}- 123.8.

Monomer III

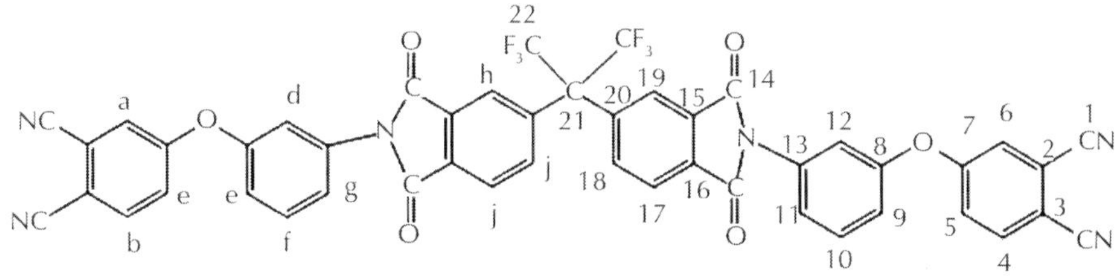

The FT-IR (KBr; cm^{-1}) 2232 (C≡N stretching), 1771 and 1725 (C=O asymm and symm stretching), 1367 (C-N stretching), 1248 (C-O-C stretching); ^1H NMR (300MHz, DMSO-d_6) δ (ppm)= 7.93 (d, 2H, H_a), 8.42 (d, 2H, H_b), 7.48 (d, 2H, H_c), 7.25–7.46 (m, 4H, $H_{d,e}$), 7.45–7.57 (d d, 2H, H_f), 7.09–7.15 (t, 2H, H_g), 7.62– 7.78 (m, 4H, $H_{h,i}$), 8.14– 8.23 (t, 2H, H_j). ^{13}C NMR (300MHz, DMSO-d_6) δ (ppm)= C_1- 116.7, C_2- 116.2, C_3- 113.8, C_4- 133.6, C_5- 122.3, C_6- 121.1, C_7- 162.4, C_8- 155.3, C_9- 118.1, C_{10}- 128.5, C_{11}- 121.9, C_{12}- 108.2, C_{13}- 133.1, C_{14}- 168.2, C_{15}- 132.1, C_{16}- 130.8, C_{17}- 130.2, C_{18}-132.1, C_{19}-126.7, C_{20}-148.8, C_{21}-68.2, C_{22}-110.2.

Monomer IV

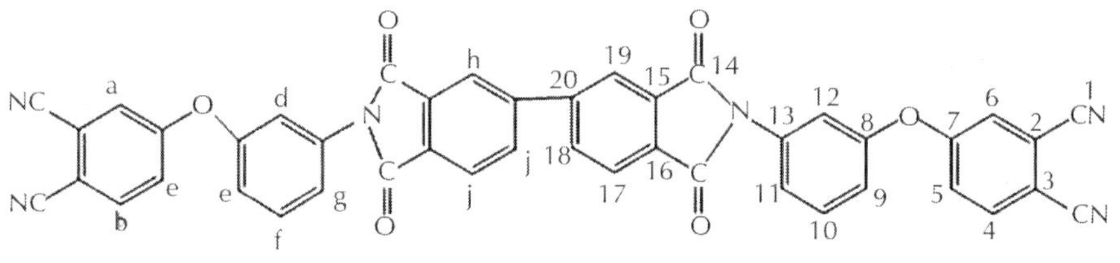

The FT-IR (KBr; cm^{-1}) 2230 (C≡N stretching), 1777 and 1726 (C=O asymm and sym stretching), 1365 (C-N stretching), 1250 (C-O-C stretching); ^1H NMR (300MHz, DMSO-d_6) δ (ppm)= 7.65–7.72 (m, 6H, $H_{a,g,c}$), 7.86 (d, 2H, H_b), 7.42 (s, 2H, H_d), 6.81 (s, 2H, H_e), 7.32 (t, 2H, H_f), 8.41 (s, 2H, H_h), 8.09 (d, 2H, H_i), 8.26 (d, 2H, H_j). The ^{13}C NMR (300MHz, DMSO-d_6) δ (ppm)= C_1- 116.2, C_2- 115.7, C_3- 114.2, C_4- 132.8, C_5- 122.1, C_6- 120.9, C_7- 161.6, C_8- 155.7, C_9- 116.1, C_{10}- 128.2, C_{11}- 120.7, C_{12}- 109.0, C_{13}- 132.9, C_{14}- 166.6, C_{15}- 132.8, C_{16}- 129.7, C_{17}-130.7, C_{18}-132.6, C_{19}-127.2, C_{20}- 146.2.

Preparation of Prepolymer Samples

All the monomers (I, II, IIIC, and IVC, see Table 1) were converted to prepolymers before the polymerization process. A typical prepolymer synthesis is as follows: In a small reaction vial 2 g of phthalonitrile monomer (IC) was heated above the melting point in air. To the melt was added (0.06 g) 3 wt % of diamine curing agent 4,4'-oxydianiline(ODA) and stirred well for about 3 min. Once the sample became homogeneous, it was quenched to room temperature. The obtained amorphous solid (B-stage resin) was ground well and used for polymerization (Laskoski et al., 2005).

Table 1. Experimental results of phthalonitrile monomers in different reaction conditions (T = 180°C).

Monomer	Time (h/m)				Yield (%)				Melting point(°C)	Cure temp (°C)
	A	B	C	D	A	B	C	D [25]	(A, B & C)	C
I	6 h	18 m	15 m	13 h	95	95	97	90	245-250	256-300
II	6 h	17 m	13 m	13 h	88	89	94	83	255-258	270-317
III	6 h	17 m	14 m	13 h	91	92	95	85	160-163	178-262
IV	6 h	17 m	15 m	13 h	89	91	95	88	237-242	258-307

A=ionic liquid; B= microwave; C= ionic liquid + microwave; D= conventional

Polymerization of Phthalonitrile Prepolymers in Microwave Irradiation

The prepared phthalonitrile prepolymers were polymerized under MW irradiation using a 3-step cure cycle. The prepolymer sample (I, II, IIIC, or IVC) (0.5 g) in a small Pyrex reaction vial was placed in MW reactor. The curing was carried out with a pre-programmed three step cycles at 260°C for 30 min, at 280°C for 30 min and at 300°C for 30 min. The temperature of the reaction was monitored using a shielded thermocouple (ATC-300) inserted directly into the corresponding reaction chamber.

RESULTS AND DISCUSSIONS

Syntheses of Phthalonitrile Monomers

The phthalonitrile terminated imide monomers were synthesized by the simple imidization procedure in three different reaction conditions using isoquinoline catalyst as shown in Scheme 1 (Keller et al., 1984). In the first method (A), the reaction was carried out in IL medium by oil bath heating technique. Initially the reaction mixture was stirred at room temperature for 3 hr under nitrogen atmosphere. Then the temperature was raised slowly to reflux for about 3 hr. The color change of the reaction content was observed during this period. The formation of the imide monomer was monitored by TLC analysis using ethyl acetate/hexane mixture. The completion of the reaction was observed by TLC. On completion of the reaction with TLC control, the resulting product was isolated by precipitation in water. The product was repeatedly washed with water and dried at 90°C under vacuum. The IL was recovered from the filtered water, kept in an oven overnight at 100°C and reused. The yield of the product was 95% (1 93% (Keller, 1993). In the second method (B), the same reaction was carried out at 180°C for about 15 min in m-cresol under MW irradiation. The product was isolated in ethanol solution and obtained by 95% yield. In method C, the reaction was carried out in IL medium under MW irradiation simultaneously. During the reaction, MWs irradiated homogeneously throughout the Teflon coated cavity by a rotating diffuser. The reaction was completed within 13 min and the product was precipitated in water. It was seen that the reaction done was obtained by the maximum yield (97%) and minimum time when compared to the other two methods (B&C).

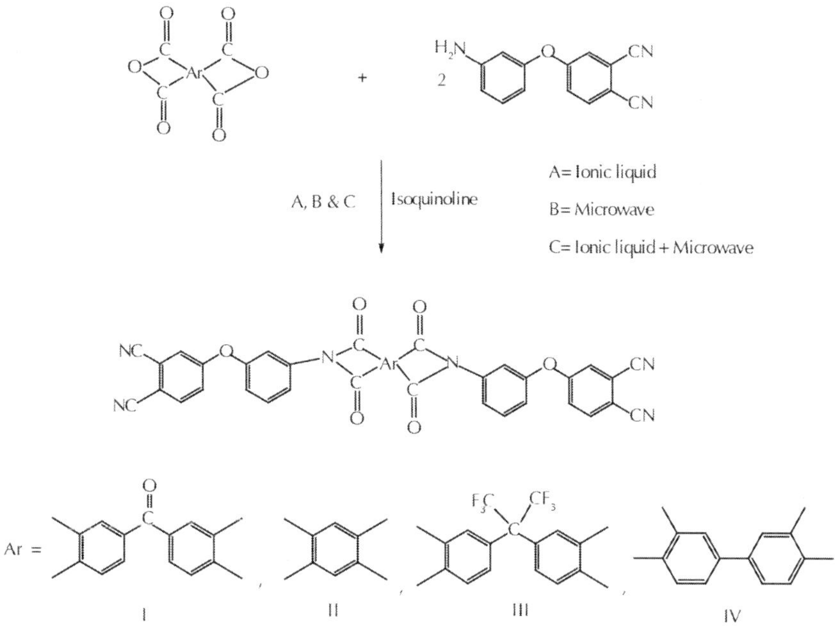

Scheme 1. Synthesis of imide-containing phthalonitrile monomers.

It has been reported that the synthesis of this monomer by the conventional method required 13 hr (Keller, 1993). With the use of MW, the total time of the reaction was reduced from hours to minutes due to the efficient and even heating, dipole relaxation of the solvent and interaction with the reactant molecules (Hoffmann et al., 2003). In addition a moderate increase (5%) in the yield is also noted. The imidization reaction under IL medium (scheme I, B) was carried out for about 6 hr and the yield was 5% higher than using the conventional organic solvent. This shows that the IL acts also as a catalyst as well as a reaction medium (Giribabu et al., 2007). The purification of the product was carried out several times and the ILs was recovered by evaporating the water. The recovered IL was heated overnight at 120°C for removing the trace of water and reused 4–5 times for the same reaction. The imidization in IL medium under MW irradiation (scheme I, C) was carried out with a constant power setting (2.45 GHz). During the reaction, TLC analyses were performed at three different time intervals for monitoring the progress and completion of the reaction. After 5 min at 180°C, a color change of the reaction mixture was observed. From the TLC results, it was observed that the imidization was completed in 13 min. The phthalonitrile monomers were precipitated and repeatedly washed with ice-cold water. The separation and removal of the product from the IL phase as the reaction proceeds allows the monomer to be obtained simply and in a highly pure state. In fact, the product contains so little of the IL that an aqueous wash step can be dispensed thoroughly. This method led to the maximum yield (97%) and the spectral analysis confirmed that the synthesized monomers were highly pure.

The structure of these compounds was confirmed by FT-IR, ^{1}H NMR and ^{13}C NMR analysis. The FT-IR spectrum of monomers I, II and III are shown in Figure 1. The characteristic absorption bands of the imide ring of monomer I were observed at 1780 cm^{-1} (C=O asymmetric stretching), 1717 cm^{-1} (C=O symmetric stretching), 1384 cm^{-1} (C–N stretching) and 724 cm^{-1} (C=O bending, out of plane). The absence of amide-carbonyl (C=O) peak at 1650 cm^{-1} confirmed the formation of the imide monomer. The absorption band corresponding to the nitrile group (C≡N) was seen at 2232 cm^{-1} and the bands at 1254 and 1071 cm^{-1} correspond to the ether (C-O-C) linkages. The ^{1}H NMR spectrum of the monomer IC is shown in Figure 2. All the proton signals were observed in the aromatic region (7.2–8.3 ppm), the meta-coupled proton signals appeared at 7.88, 7.89 ppm as a doublet and ortho-coupled proton signal appeared at 8.14, 8.15 ppm. The *ortho* and *meta* coupled proton signals appeared as two doublets at 7.51, 7.52, 7.49, 7.48 ppm. The rest of the proton signals confirmed the proposed structure (Kumar et al., 1993).

Figure 3 shows the ^{13}C NMR spectrum of phthalonitrile monomer IC in DMSO-d_6. The bridging carbonyl group signal appeared at 194 ppm and the carbon signals at 154.3, 161.1, 166.9 ppm due to the presence of carbonyl and phthalonitrile group. The spectral assignments are presented in the experimental section (3.4).

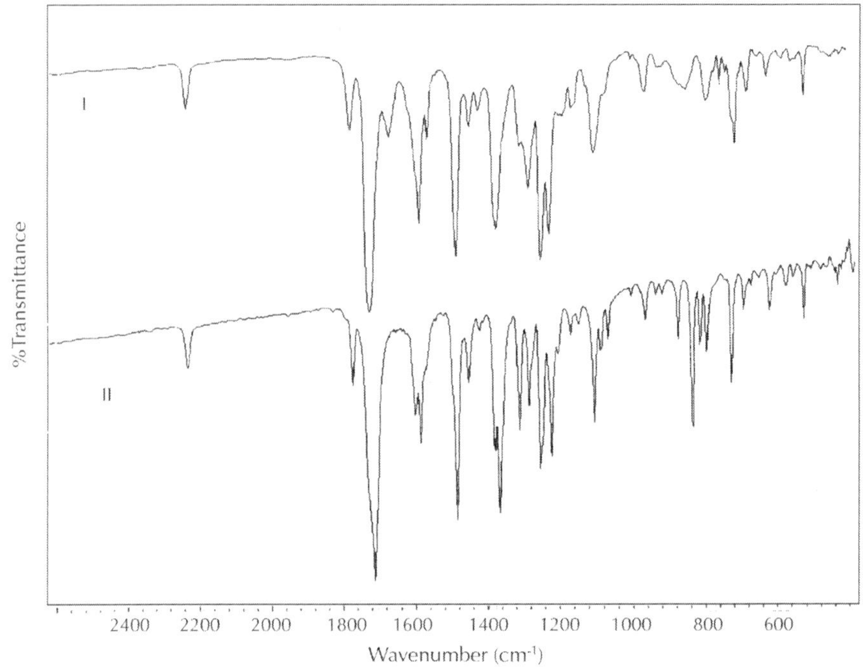

Figure 1. FT-IR spectrum of phthalonitrile monomer I (top) and II (bottom).

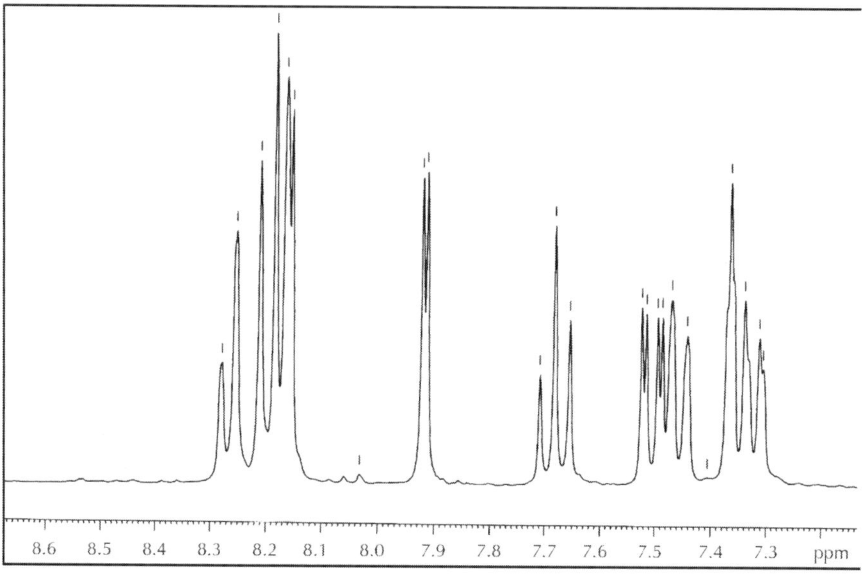

Figure 2. ^{1}H NMR spectrum of phthalonitrile monomer IC.

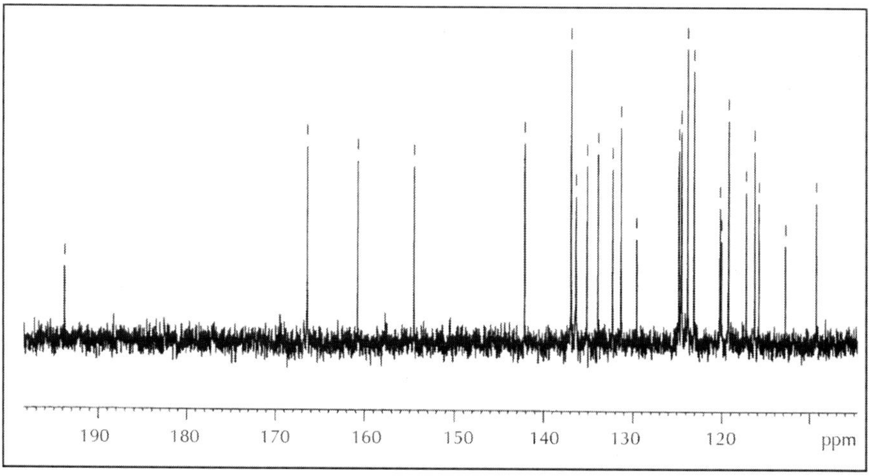

Figure 3. ^{13}C NMR spectrum of phthalonitrile monomer IC.

Differential Scanning Calorimetry

The melting and polymerization temperatures of the monomers were analyzed by differential scanning colorimeter (DSC) under nitrogen atmosphere and the results are summarized in Table 1. Monomer IC with 3 wt % of diamine (ODA) curing agent shows that the onset and completion temperatures of polymerization reaction are 256 and 300°C, respectively (l260–315°C and the reaction time of 32–36 hr) The same monomer was cured under MW irradiation between 252 and 300°C for about 90 min in a three-step cycle (260°C/30min, 280°C/30min, 300°C/30min). The FT-IR spectrum (Figure 4, spectrum I) shows that after 60 min in MW irradiation almost all the nitrile groups are cross-linked as networks.

A trace amount of nitrile groups remained as unreacted due the increase of viscosity which decreases the chain mobility. After 90 min, there is no evidence of nitrile groups (Figure 4, spectrum II). The above results confirm that MW irradiation significantly increases the rate of the reaction and hence the reaction time can be reduced.

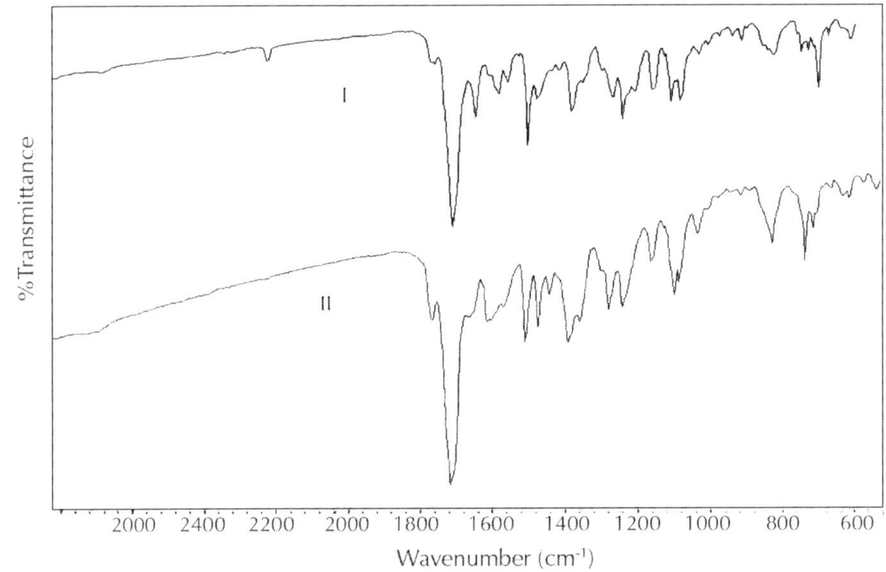

Figure 4. FT-IR spectrum of phthalonitrile prepolymer IC: (I) after 60 min cure with 3 wt % amine (ODA) under microwave; (II) prepolymer IC after curing at 300°C for 30 min under microwave.

Polymerization under Microwave (MW) Irradiation

The conventional curing of phthalonitrile resin system has been shown to proceed very slowly even during extended of time at elevated temperatures. Polymerization under MW irradiation allowed the cross-linking reaction to proceed faster and at a fairly low temperature. The phthalonitrile monomers (I, II, III, and IVC) were converted to prepolymers by adding 3 wt % of 4,4′-oxydianiline (ODA) curing agent (Scheme 2). The prepolymer preparation is the preliminary step of the cure reaction. The advantages of the prepolymer synthesis are to decrease the viscosity of the polymerization mixture and enhance the processability. The amorphous prepolymers were cured as crosslinked network under MW with a three step cure cycle. The cure reaction takes place through the terminal cyano group of the monomers. Polymerization was achieved for about 90 min under MW irradiation with 3 wt % of diamine (ODA) curing agent.

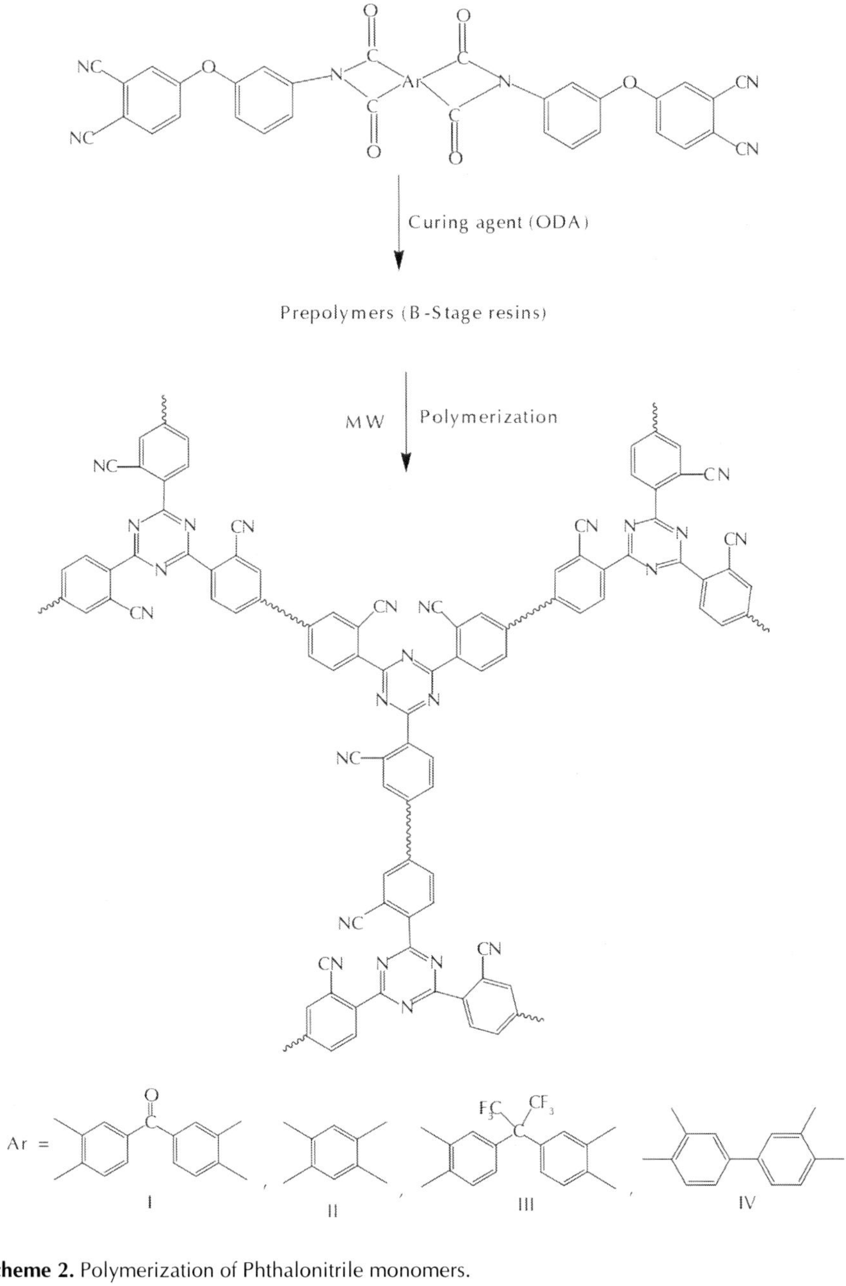

Scheme 2. Polymerization of Phthalonitrile monomers.

The FT-IR analysis was used to monitor the progress of cure reaction under MW irradiation. Figure 4 shows the FT-IR spectrum of prepolymer (IC) and cured resin. With the first 30 min of polymerization at 260°C under MW irradiation the intensity of the nitrile absorption band at 2,232 cm^{-1} decreased and new bands appeared at 1,395 and 1,520 cm^{-1}. This confirmed that the cross-linking process started at the cynano group through the formation of triazine ring structures (Selvakumar et al., 2009). Then the reaction was continued at 280°C for 30 min and at 300°C for 30 min. After 90 min, the FT-IR analysis showed that almost all the nitrile groups were converted into cross-linked networks except very small nitrile (C≡N) absorption (Sastri et al., 1998, 1999). The results obtained from all the methods are summarized in Table 1.

CONCLUSION

Imide-containing phthalonitrile monomers were successfully synthesized by using the IL 1-methyl-3-imidazolium chloride under MW irradiation with high yield and purity. The MW heating is a very efficient energy source and can be used to dramatically reduce reaction time and significantly increase the yield. The synthesized monomers were cured with 3 wt % of curing agent (ODA) under MW condition and compared to the conventional synthesis techniques. The results show that the IL is a very good candidate to replace the volatile organic solvents. The MW irradiation strongly enhanced the rate of polymerization and processability.

KEYWORDS

- **Ionic liquids**
- **Microwave**
- **Phthalonitrile**

ACKNOWLEDGMENTS

This work was supported by a seed grant from the Shastri Indo-Canadian Institute, Calgary, Canada and by Natural Sciences and Engineering Research Council of Canada.

Chapter 5

Nanocomposite Polymer Electrolytes for Lithium Batteries

N. Angulakshmi, K. S. Nahm, V. Swaminathan, Sabu Thomas, and R. Nimma Elizabeth

INTRODUCTION

The physical and electrochemical properties of a new class of lithium ion conducting polymer electrolytes formed by dispersing nanosized $Ca_3(PO_4)_2$ in the poly (vinylidene fluoride-hexafluoro propylene) (PVDF-HFP) – $LiClO_4$ complexes have been reported. The prepared membranes were subjected to XRD, SEM, TG-DTA, and FT-IR analysis. Ionic conductivity measurements have been made as a function of temperature and lithium salt concentrations. The polymeric film with a ratio of PVDF-HFP:$Ca_3(PO_4)_2$:$LiClO_4$; 75:15:10 offered maximum ionic conductivity. The interfacial property of Li/NCPE was also analyzed. The interaction that exists between the polymer and lithium salt species has been confirmed by FT-IR analysis.

The forecast of energy consumption relying on fossil fuels to cause a severe problem in the world economics and environmental concerns and the growing market of consumer electronic products have motivated the research and development of electrochemical power sources (Gray, 1997; Manuel Stephan, 2006; McCallum et al., 1987). These electrochemical power sources are expected to meet out high energy/power densities, good cycle life, reliability, and safety. Rechargeable lithium battery with lithium metal anode has been identified as the ultimate candidate due to its high gravimetric and power densities (Gray, 1991). However, high reactivity lithium metal anode with liquid electrolytes leads to the formation of surface layers and other uncontrolled phenomena.

This lithium electrode/electrolyte interfacial problem can be circumvented by replacing liquid organic electrolyte with a solid state, solvent free lithium polymer electrolyte (Scrosati, 1993).

Poly (ethylene oxide) (PEO) – LiX complexes appear to be the most suitable electrolytes for lithium polymer batteries, however, the local relaxation and segmental motion of the polymer chains remain a problem area (Armand et al., 1997). Therefore, the PEO-based electrolytes show an appreciable ionic conductivity only above 100°C (Gorecki et al., 1986). This is, of course, a drawback for applications in the consumer electronic market. On the other hand, the gel polymer electrolytes although offer high ionic conductivity and appreciable lithium transport properties it suffers from poor mechanical strength and interfacial properties (Croce et al., 1998; Gray et al., 1986; Kelly et al., 1985; Weston et al., 1982). Recent studies reveal that the nanocomposite polymer electrolytes alone can offer safe and reliable lithium batteries (Appetecchi et al., 2000; Itoh et al., 2003; Wieczorek, 1992). However, the ionic conductivity of nanocomposite polymer electrolytes is found to be low at ambient and sub-ambient temperatures (Capuoano et al., 1991). In order to enhance the ionic conductivity at ambient temperature, numerous research efforts have been dedicated (Manuel et al., 2006). The most successful work has been incorporation of inorganic materials such as ceramic and nano-oxides, layered clays, and so on (Appetecchi et al., 2000; Capuoano et al., 1991; Croce et al., 1998; Itoh et al., 2003; Weston et al., 1982; Wieczorek, 1992).

Recent studies also reveal that nanosized fillers are more effective in enhancing the ionic conductivity, lithium ion transference number and electrochemical stability (Kumar et al., 1994). Wang et al. (2010) introduced a fast ionic conductor with $Li_{1.3}AlSO_{.3}Ti_{1.7}(PO_4)_3$ as filler in the PEO matrix. The authors achieved a high ionic conductivity of 4.53 x 10^{-4} S/cm at room temperature.

In the recent past PVDF-HFP has been identified as a potential candidate due to its appealing properties; the polymer host itself has a dielectric constant of 8.4 which aids higher dissociation ionic species. Also the PVDF—crystalline phase acts as mechanical support while HFP—amorphous phase facilitates for higher ionic conduction (Song et al., 1999). Recently, Li et.al. (2008) briefly reviewed the preparation, physical and electrochemical properties of PVDF-HFP based gel polymer electrolytes. Also nano $Ca_3(PO_4)_2$ has been used as filler for the first time. Shin et al. (2005) and Appetecchi et al. (2003) have shown that membranes prepared by conventional solvent casting method lead to poor interfacial properties, at the Li-polymer interfaces and the traces of solvent react in the high surface area, nanosized inert fillers even after prolonged drying. Hence, in the present study the hot-press technique is employed for preparation of nanocomposite polymer electrolytes.

EXPERIMENTAL

Preparation of Nano $Ca_3(PO_4)_2$

The nanosized calcium phosphate filler particles were synthesized using in situ deposition technique in the presence of PEO as reported by one of the authors (Thomas et al., 2009). Firstly, a complex of calcium chloride with PEO (Aldrich, USA) was prepared in desired proportions in methanol. An appropriate stoichiometric amount of trisodium phosphate, $Na_3(PO_4)$ in distilled water was added to the above complex slowly without stirring. The whole mixture was allowed to digest at room temperature for 24 hr when both the chloride and phosphate ions diffused through the PEO and formed a white gel like Precipitate, which was filtered, washed and dried. The samples were prepared for different molar concentrations of PEO-$CaCl_2$ complex of 2:1, 4:1, and 5:1 and the yield of the calcium phosphate was recorded as 83, 75, and 67%. In order to remove excess PEO and NaCl the prepared sample was washed with double distilled water several times until it reaches the pH value of 7. Figure 1 shows the TEM images of $Ca_3(PO_4)_2$ prepared with different compositions of PEO. However, in the present chapter the sample prepared with the ratio of PEO:$CaCl_2$; 4:1 was used. The average particle size of the synthesized $Ca_3(PO_4)_2$ particles was found to be less than 20 nm.

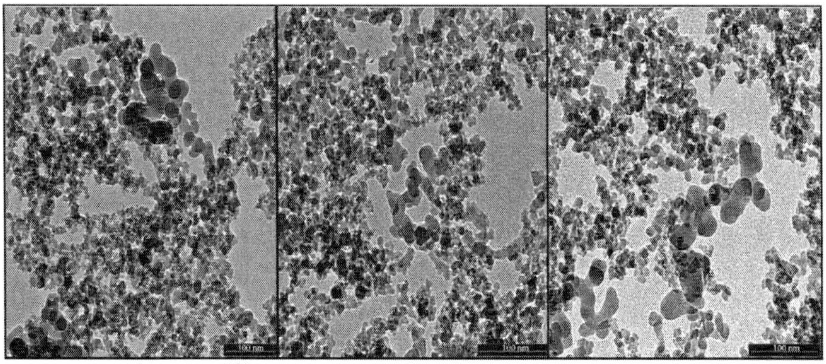

Figure 1. TEM image of the calcium phosphate nanoparticles (PEO: $CaCl_2$, (a) 2:1, (b) 4:1, (c) 5:1).

Preparation of Nanocomposite Polymer Electrolytes

The PVDF-HFP (Aldrich, USA) and lithium perchlorate, $LiClO_4$ (Merck, Germany) were dried under vacuum for 2 days at 50 and 100°C, respectively. The $Ca_3(PO_4)_2$ was also dried under vacuum at 50°C for 5 days before use. Nanocomposite electrolytes were prepared by dispersing appropriate amounts of $Ca_3(PO_4)_2$ in PVDF-HFP for different concentrations of $LiClO_4$ (as depicted in Table 1) and the powder was hot-pressed into films, as described elsewhere (Appetecchi et al., 2003; Manuel et al., 2009). The nanocomposite electrolyte films had an average thickness of 30–50 μm. This procedure yielded a homogenous and mechanically strong membrane, which were dried under vacuum at 50°C for 24 hr for further characterization.

Table 1. Composition of PVdF-HFP, $Ca_3(PO_4)_2$ and $LiClO_4$.

Sample	PVdF-HFP Wt.%	$Ca_3(PO_4)_2$ Wt.%	$LiClO_4$ Wt.%
S1	95	0	5
S2	90	5	5
S3	85	10	5
S4	75	17	8
S5	70	20	10
S6	94	5	1
S7	93	5	2
S8	92	5	3
S9	91	5	4

Electrochemical Characterization

The ionic conductivity of the membranes sandwiched between two stainless steel blocking electrodes (1 cm^2 diameter) was measured using an electrochemical impedance analyzer (IM6-Bio Analytical Systems) with frequency range from 50 mHz to 100 kHz at various temperatures (0, 15, 30, 45, 60, 70, and 80°C).

Symmetric non-blocking cells of the type Li/CPE/Li were assembled for compatibility studies and were investigated by studying the time-dependence of the impedance of the systems under open-circuit at 60°C.

Morphological examination of the films was made by a scanning electron microscope (FE-SEM, S-4700, Hitachi) under a vacuum condition (10^{-1} Pa) after sputtering gold on one side of the films. Differential scanning calorimetry measurements were performed at a rate of 10°C min^{-1} between 20 and 250°C while TG-DTA in the temperature range 20–300°C. The lithium/polymer electrolyte interface was analyzed using FTIR (Thermo NICOLET Corporation, Nexus Model -670) by single internal reflection (SIR) mode (Chusid et al., 2001). The infra-red spectra were obtained at ambient temperature with an 8 cm^{-1} resolution.

RESULTS AND DISCUSSION

XRD Analysis

X-ray diffraction analysis is very useful in determining the structure of materials. X-ray diffraction patterns of samples PVDF-HFP, $Ca_3(PO_4)_2$, PVdF-HFP + $LiClO_4$(S1) and PVdF-HFP + $LiClO_4$ + $Ca_3(PO_4)_2$, (S5) are respectively depicted in Figure 2(a–d). The diffraction peaks at $2\theta = 18.4$, 20, and 26.6 correspond to (1 0 0) (0 2 0), and (1 1 0) reflections of crystalline PVDF (Chusid et al., 2001).This confirms partial crystallization of the PVDF units in the copolymer to give an over-all semi-crystalline morphology for PVDF-HFP (Abbernt et al., 2001). Since $LiClO_4$ is complexed in the polymer matrix, peaks corresponding to $LiClO_4$ are not observed which also indicate that the lithium salt is completely dissolved in the polymer matrix (Figure 2(c)). Figure 2(d)

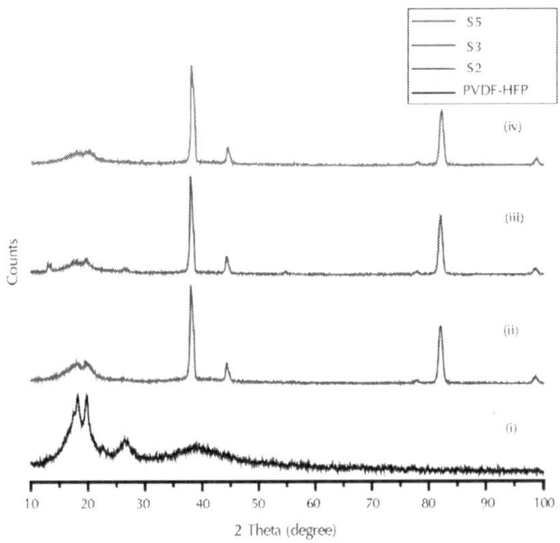

Figure 2. X-ray diffraction of (i) PVdF-HFP (ii) sample S2 with $Ca_3(PO_4)_2$ (5%) (iii) sample S3 with $Ca_3(PO_4)_2$ (10%) (iv) sample S5 with $Ca_3(PO_4)_2$ (20%).

shows that the intensity of crystalline peaks decreased and broadened upon incorporation of $Ca_3(PO_4)_2$. However, the $Ca_3(PO_4)_2$ peaks remain unaltered. These observations apparently reveal that the polymer undergoes significant structural reorganization while adding inert fillers and salts. Upon the addition of nano $Ca_3(PO_4)_2$ and lithium salt the crystallinity of the polymer has been considerably reduced. This reduction in crystallinity is attributed to small particles of inert filler, which changes the chain reorganization and facilitates higher ionic conduction (Wieczorek et al., 1996).

Thermal Analysis

In order to evaluate the thermal stability of polymer electrolytes the thermogravimetric analysis of the NCPE was performed under nitrogen atmosphere. The TG-DTA traces of PVDF-HFP + $LiClO_4$(S1) and PVDF-HFP + $Ca_3(PO_4)_2$ + $LiClO_4$(S4) membranes are displayed in Figure 3(a) and (b) respectively. A weight loss of about 3% was observed about 116°C for sample (S1) and is attributed to the release of water absorbed at the time of loading the sample (Ribeiro et al., 2001; Shodai et al., 1994). Furthermore, no weight loss was observed up to 304°C which indicates that no decomposition was induced by the thermal steps of the hot press process (Chusid et al., 2001). A peak exhibited in DTA curve at about 140°C is attributed to the eutectic transition in the electrolytes (Shodai et al., 1994). Upon addition of $Ca_3(PO_4)_2$ in the polymer matrix Figure 3(b) the decomposition temperature is to advanced 260°C which is attributed to the decomposition of $Ca_3(PO_4)_2$ particles.

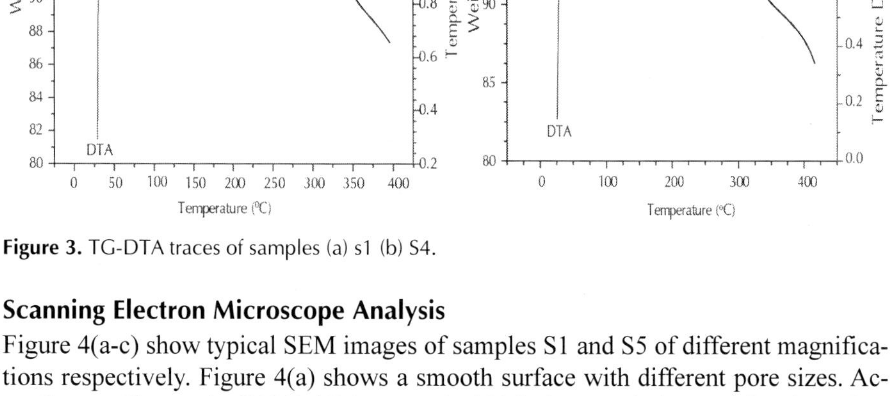

Figure 3. TG-DTA traces of samples (a) s1 (b) S4.

Scanning Electron Microscope Analysis

Figure 4(a-c) show typical SEM images of samples S1 and S5 of different magnifications respectively. Figure 4(a) shows a smooth surface with different pore sizes. According to Chu et al., (2003) (Chiang et al., 2004) the morphology of the electrolyte surface can be modified/tailored by the incorporation of both ionic salts and fillers. The SEM image (Figure 4(b)) of PVDF-HFP + 20% $Ca_3(PO_4)_2$ + $LiClO_4$ (sample S5) shows a relatively rough surface with inhomogeneous morphology. Although it was intended to prepare a uniform distribution of nano $Ca_3(PO_4)_2$ in PVDF matrix the aggregation of nanoparticles with increasing concentration of nano $Ca_3(PO_4)_2$ could not be achieved in the porous PVDF-HFP based electrolytes, as evidenced in Figure 4(c) (PVdF-HFP + $LiClO_4$ + 20% $Ca_3(PO_4)_2$ sample S5) at high contents of nano $Ca_3(PO_4)_2$ (20%). The membrane is seen to have a rough surface with an inhomogeneous morphology with islands of aggregated particles on the surface of the polymeric membrane (Chiang et al., 2004).

Ionic Conductivity

Figure 5(a) and (b) shows the variation of ionic conductivity as a function of temperature for different concentrations of lithium salt and nano $Ca_3(PO_4)_2$ respectively. It is well known that Li ions migrate in two ways: (a) move along the molecular chains of polymer, and (b) move in the amorphous phase of polymer electrolyte (Robitailla et al., 1986). The former is slow transport where as the latter is fast. Figure 5(a) illustrates that the ionic conductivity increases with the increase of temperature and salt concentration. The ionic conductivity gradually increases with the increase in $LiClO_4$ content (S6–S9) and also increases with the increase in temperature up to 70°C and at the 80°C, the ionic conductivity of S6, S7, and S8 are lower than the S1 but S9 is higher than the S1. These results are in accordance with those reported on PEO-based polymer electrolytes with SiO_2—lithium imide anion system (Wang et al., 2003). It can be seen from Figure 5(b), the ionic conductivity increases with increase of $Ca_3(PO_4)_2$ in the polymer film. The competition between $Ca_3(PO_4)_2$ and F atoms with respect to lithium ions facilitates the migration for lithium ions in polymer electrolyte and hence the ionic conductivity of polymer electrolyte increases with the addition of $Ca_3(PO_4)_2$ into polymer matrix.

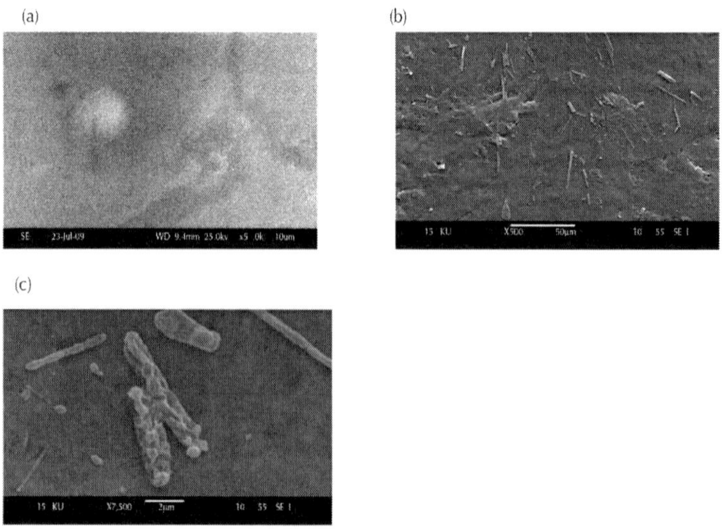

Figure 4. Scanning Electron Microscope images of (a) PVdF-HFP + $LiClO_4$ and (b) and (c) sample of S5 with different magnification.

When the content of $Ca_3(PO_4)_2$ in the NCPE is increased to 20% the ionic conductivity of the NCPE decreases. This decrease in the ionic conductivity can also be attributed to the change in the crystallinity of PEO in the nanocomposite polymer electrolytes (Capuglia et al., 1999). According to Scrosati and co-workers (Scrosati et al., 2001), the Lewis acid groups of the added inert filler may compete with the Lewis acid lithium cations for the formation of complexes with the PEO chains as well as the anions of the added lithium salt. In the present study, the filler nano $Ca_3(PO_4)_2$, which has a basic center can react with the Lewis acid centers of the polymer chain and these interactions lead to the reduction in the crystallinity of the polymer host. Nevertheless, the result provides Li^+ conducting pathways at the filler surface and enhances ionic transport.

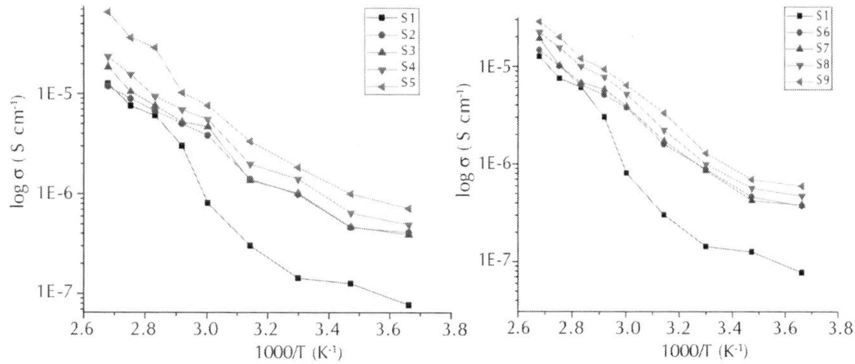

Figure 5. Ionic Conductivity as a function of temperature for the various (a) $Ca_3(PO_4)_2$ concentrations and (b) salt concentrations.

FTIR Analysis

The FTIR has been used as a powerful to characterize the molecular and structural changes in the polymer electrolyte systems. Figure 6(a–d) show the FTIR spectra of pure PVDF-HFP, $Ca_3(PO_4)_2$, PVDF-HFP + $LiClO_4$ and PVDF-HFP + $LiClO_4$ + $Ca_3(PO_4)_2$ (sample S5). The vibrational bands at 490 cm^{-1}(Figure 6(a)) are assigned to the wagging vibrations of CF_2 group. The frequencies of 1071, 759, and 607 cm^{-1} belongs to the vibration of crystalline phase of (PVDF-HFP) whereas the frequencies of 883 and 841 cm^{-1} are assigned to the vibration of amorphous phase of (PVDF-HFP) (Abbernt et al., 2001). The CH_2 ring and CH groups absorb in the regions of 3100–2990 cm^{-1}. The peaks appearing at 1,192 and 2,923 cm^{-1} are assigned to the asymmetrical stretching vibrations of CF_2 group and CH_2 group respectively (Rajendran et al., 2002). The deformation vibration of CH_2 group which appears at the frequency 1,402 cm^{-1} (Pearson et al., 1960) will move to a higher frequency/position with the weakening of interaction between H atoms of CH_2 groups and F atoms of CF_2 groups. The vibration bands at 1,647 and 3,418 cm^{-1} can be assigned to the stretching and bending vibrations of OH bonds of the absorbed water. Upon incorporation of $LiClO_4$ in polymer host, the peak at 758 cm^{-1} shift to 748 cm^{-1} attributed to wagging band of CH_2Cl. Frequencies

840–560 cm^{-1} are assigned to C-Cl stretching vibrations. The characteristic absorption vibrations of $LiClO_4$: 1,150–1,080 cm^{-1}, 941 cm^{-1} are assigned to symmetrical vibration of ionic pairs between Li^+ and ClO_4^- (Focher et al., 1992), 624 cm^{-1} is stretching vibration of ClO_4^- and the sharp peaks 3,597 and 1,637 cm^{-1} are stretching and bending vibrations of OH bonds of the absorbed water. The absorption peaks at 2,983 and 3,025 cm^{-1}, which are assigned to the symmetrical and non-symmetrical stretching vibration of CH_2 groups, appear after addition of $LiClO_4$ to polymer film, because of the interaction between lithium ions and F atoms. It is found that the intensity of these two absorptions decreases with the addition of nano $Ca_3(PO_4)_2$ (Figure 6(d)) in the polymer matrix. Figure 6(d), shows that the deformed vibration of CH_2 group moves to 1,407 cm^{-1} after addition of nano $Ca_3(PO_4)_2$ into polymer matrix from 1,402 cm^{-1}.

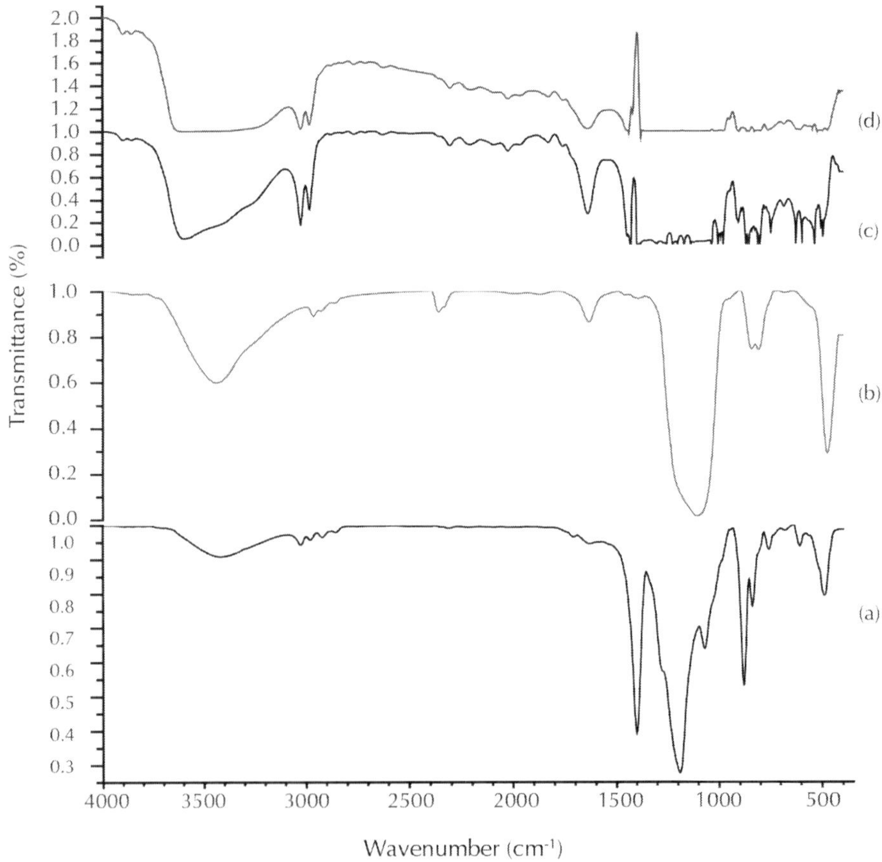

Figure 6. FTIR Spectra of (a) PVDF-HFP (b) Ca3(PO4)2 (c) sample S1 and (d) sample S5.

Interfacial Property of Li/NCPE

Figure 7(a) shows the FTIR spectrum of NCPE (Sample S5) consisting PVDF-HFP + $LiClO_4$ + $Ca_3(PO_4)_2$. The FTIR spectra obtained on a lithium electrode in contact with PVDF-HFP + $LiClO_4$ and PVDF-HFP + $LiClO_4$ + $Ca_3(PO_4)_2$ polymer membranes (through KBr window) are displayed in Figure 7(b) and (c) respectively. In Figure 7(c), the original features belonging to the polymer electrolytes not only disappeared, but new peaks have arisen. When a nano $Ca_3(PO_4)_2$ incorporated composite electrolyte comes in contact with the lithium metal anode, new peaks appear at 1,754, 1,579, 1,492, 938, and 405 cm^{-1}. The peak that appears at 405 cm^{-1} is attributed to Li-N and 938 cm^{-1} is assigned to unconjugated cyclic anhydrides. The peaks at 1,754, 1,579, and 1,491 cm^{-1} are attributed to C=O, CH_2 stretching and C-C skeletal vibrations. (Huang et al., 1996) It is significant that no predominant peak appears around 1,100 cm^{-1} and is attributed to the fact that $LiClO_4$ is reduced to $LiClO_4$ species thus confirming its presence (Schechter et al., 1999).

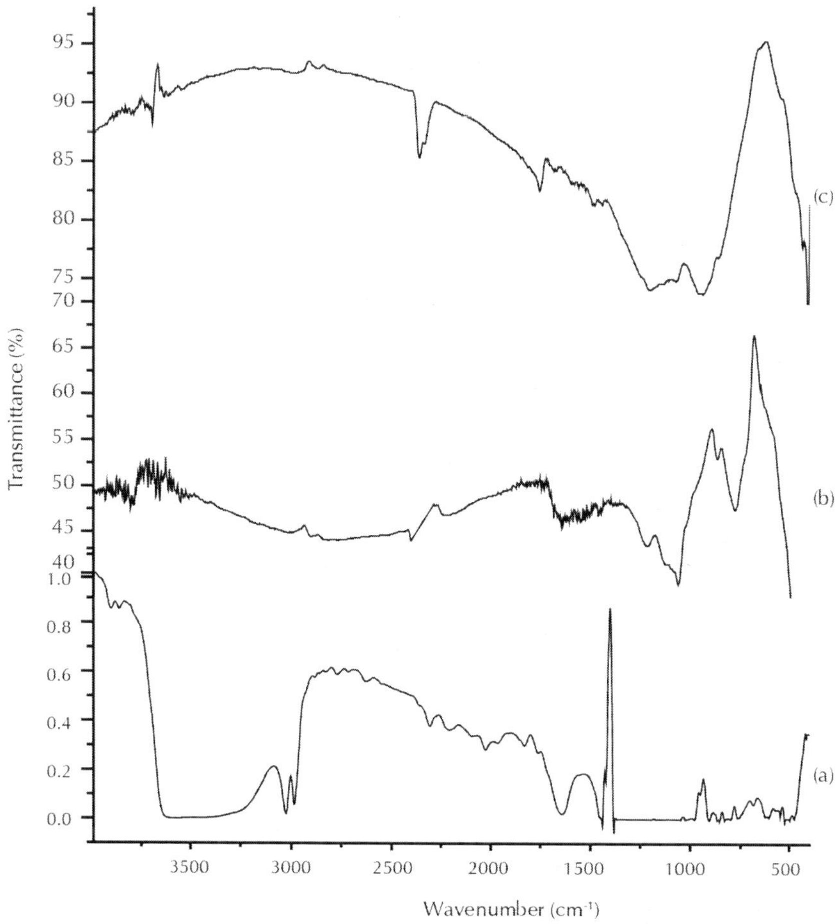

Figure 7. FTIR spectra of (a) NCPE (S5) (b) Li-PE (S1) and (c) Li-NCPE (S5).

Tensile Measurements

In order to quantify the mechanical strength, the stress-strain properties have been measured and are shown in Figure 8. The stress-strain behavior gives fundamental information about the load vs. deformation characteristics of polymer composites. In the case of unfilled system (0% $Ca_3(PO_4)_2$), the sample shows two distinct regions: the initial linear region reflects the elastic characteristic and the non-linear region shows the plastic deformation. As expected the neat sample (sample S1) has high elongation-at-break as compared to the filled nanocomposites except for the sample S5 (Fan et al., 2006). Upon the addition of $Ca_3(PO_4)_2$ we can see that the moduli and tensile strength increase with dramatic decrease in strain at break. The maximum tensile strength and moduli are obtained at 10% addition of $Ca_3(PO_4)_2$ (S3). At very high concentration of $Ca_3(PO_4)_2$ (>10%) these properties decrease an account of the aggregation of $Ca_3(PO_4)_2$ particles. The high elongation at break of S5 (20% of $Ca_3(PO_4)_2$) clearly shows poor polymer/filler interaction due to agglomeration.

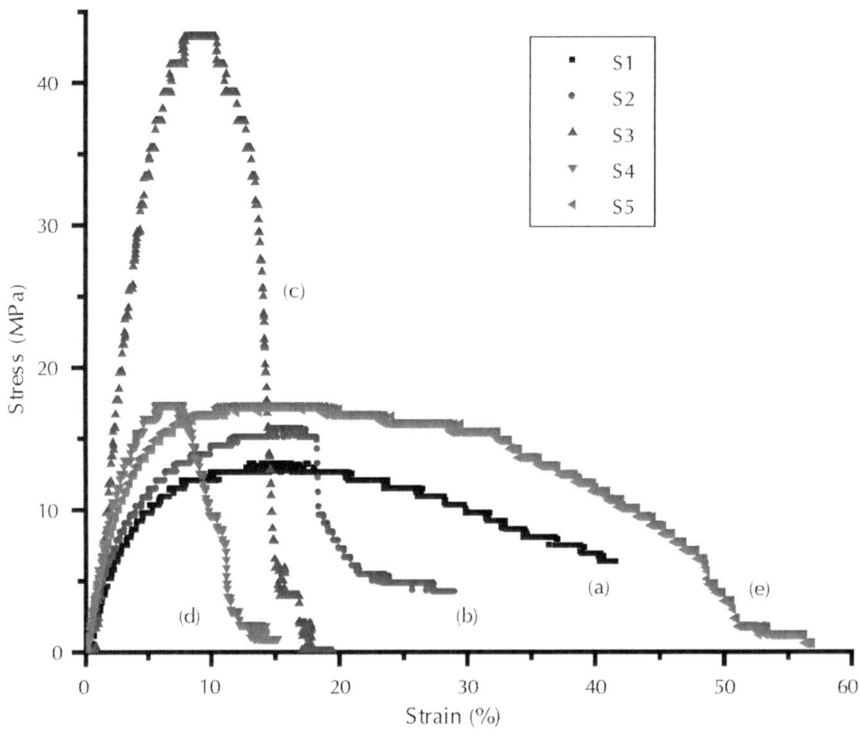

Figure 8. Stress vs. Strain curves of polymer electrolytes PVDF-HFP+$LiClO_4$+ x% Ca3(PO4)2 (a) x = 0, (b) x = 5, (c) x = 10, (d) x = 17, and (e) x = 20.

CONCLUSION

The PVDF-HFP based NCPE electrolytes were prepared for various concentrations of nano $Ca_3(PO_4)_2$ and $LiClO_4$. The nanocomposite polymer electrolytes with a composition of PVdF-HFP: $Ca_3(PO_4)_2$: $LiClO_4$ (80:15:5) exhibited maximum ionic conductivity at 80°C. The maximum tensile strength and moduli are obtained at 5 % nano $Ca_3(PO_4)_2$ - added polymer electrolytes. These electrolytes are found to be stable thermally up to 260°C which is much higher than the operating temperature of normal lithium batteries. The addition of $Ca_3(PO_4)_2$ up to 10 % of $Ca_3(PO_4)_2$ dramatically increases the moduli and tensile strength with a decrease in strain at break.

KEYWORDS

- **Ionic conductivity**
- **Lithium batteries**
- **Polymer electrolytes**
- **Polymer electrolytes/Li interfacial properties**

Chapter 6

Crack Growth Rates of Rubber Vulcanizates

M. Bijarimi, Azemi Samsuri, and H. Zulkafli

INTERODUCTION

A new method was developed for the measurement of rubber cyclic crack growth. The effects of crack growth rates on carbon black filler loadings and polymer type were investigated. The crack growth rate has been found to be lower, at higher concentration of carbon black loading in NR vulcanizates. The effect on crack growth has been attributed to pronounced changes induced by carbon black on stiffness and hysteresis properties. This mechanism is responsible for the increased resistance to crack growth. As for the polymer type, the crack growth resistance was found to be in the ranking order of NR > NBR > SBR > BR due to the influence of glass transition temperature, T_g.

Rubber is an ideal material for most of engineering products such as tyres, seals, engine mounts, bushes, belts and hoses. In the service application, high crack growth resistance is desired to minimize premature failure associated with dynamic or cyclic induced cracking. Currently, the most widely known and standardized apparatus for crack growth test is the De Mattia according to ISO 132 and ISO 133 test methods. However, the interpretation of the De Mattia test results is very subjective and operator dependent. As such the fracture mechanics approach is widely used to characterize the crack growth resistance of rubber vulcanizates. The most common test pieces to determine the tearing energy or the crack growth resistance of vulcanized rubber are trouser tear, pure shear, parallel strips with edge and angled type test pieces. The tearing energy theory developed by Rivlin and Thomas (Lake et al., 1964a; Rivlin et al., 1953; Thomas, 1958) has been used successfully in various types of fracture of rubber like materials such as tearing, fatigue, abrasion, cutting by sharp object and cyclic crack growth.

The use of conventional test pieces such as trousers, pure shear, angled and edge crack (Figure 1) in determining the crack growth of gum vulcanizates is not an issue because the crack propagates along the intended path. However, the phenomenon of forking or deviation of the crack tip from the intended tear path is common for the black-filled vulcanizates. When forking occurs at the crack tip, interpretation of the results is made difficult since the initial single crack tip branches into two or three new tips. In order to circumvent this problem, split-tear test piece (Figure 2) is preferred to other test pieces. Split-tear test was originally introduced by Thomas (Thomas, 1960) and later used by Azemi et al. (1988) to investigate the effect of strength anisotropy on the development of knotty tear. In the present study, we have developed a prototype machine that is able to measure the crack growth resistance under cyclic loading (Azemi et al., 2004). This chapter reports the study of the crack growth rates of carbon black-filled loading and different types of polymer.

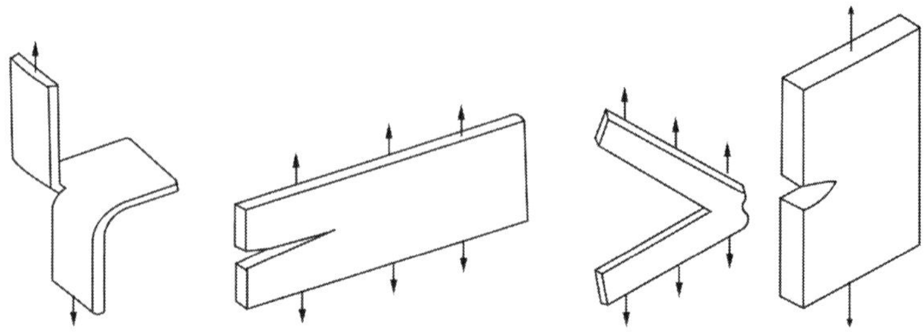

Figure 1. Conventional test piece geometries.

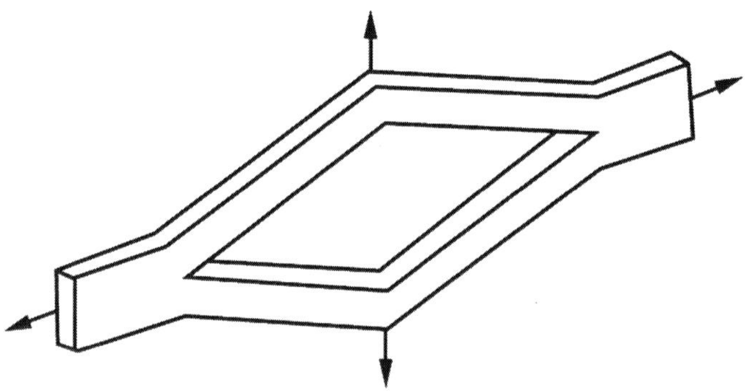

Figure 2. Split-tear test piece.

EXPERIMENTAL

Materials

Rubber mixes were prepared in the Banbury mixer according to the recipes in Table 1–2. The rheological properties of all compounds were determined by using an oscillating disc Monsanto Rheometer MDR 2000 and Mosanto Viscometer VM2000.

Table 1. Formulations for effect of different polymers.

Ingredients	SBR	NBR	BR
SBR	100	–	–
NBR	–	100	–
BR	–	–	100
Zinc oxide	5	5	5
Stearic acid	1.5	1.5	1.5
ISAF (N220)	50	50	50
IPPD	1	1	1
Sulfur	2	2	2
CBS	0.5	0.5	0.5
TMTM	0.7	0.7	0.7
Total	160.7	160.7	160.7

Table 2. Formulations for effect of different carbon black filler loading,

Ingredients	30phr	40phr	60phr
NR	100	100	100
Zinc oxide	5	5	5
Stearic acid	1.5	1.5	1.5
ISAF (N220)	30	40	60
IPPD	1	1	1
Sulfur	2	2	2
CBS	0.5	0.5	0.5
TMTM	0.7	0.7	0.7
Total	140.7	150.7	170.7

Cyclic Crack Growth

Split-tear test pieces were prepared by die stamping on a thin flat sheet of vulcanized rubber of uniform thickness. The test piece was clamped by a grip, and a dead load F_B applied to the other end of the test piece via a frictionless pulley. A dead load, F_A which was attached to a straight cylindrical iron rod was applied on each side of the test piece via a frictionless pulley in the transverse direction to F_B. The total load, F_A was the sum of the dead load and the weight of the cylindrical rod. The opening and closing of the cut contributed to cyclic crack growth. The crack grew in the direction of F_B. The rate of opening and closing of the cut (its frequency) could be regulated by choosing the desired motor speed. The crack lengths were measured with an eye-piece lens scale after a certain number of cycles. It was necessary to plot crack length measured in the unstrained state, c_o, versus number of cycles, N, at each tearing energy value. From the slope of the straight line, crack-growth cycle (dc/dN) could be obtained as shown in Figure 1(b). The tearing energy for the split-tear test piece can be computed using the equation (1) given below (Thomas, 1960). By varying loads F_A and $F_{B,}$ series of tearing energy values could be obtained from which a plot of tearing energy versus dc/dN can be produced.

$$T = \lambda\,[(F_A^{\,2} + F_B^{\,2})^{1/2} - F_A]/t \qquad (1)$$

where λ = average extension ratio in simple extension regions A and B.
t = average nominal thickness of the test piece
T = tearing energy

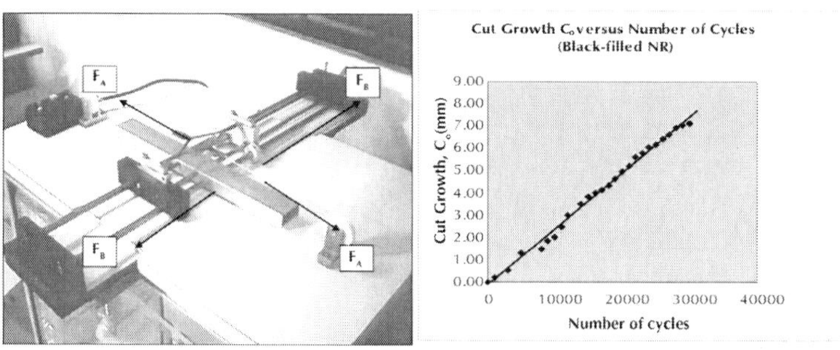

(a) (b)

Figure 3. (a) Forces F_A and F_B acting on the split-tear test piece (b) Crack length vs. number of cycles.

RESULT AND DISCUSSIONS

Effect of Polymer Type

At the same tearing energy about 0.8 kJ/m^2, the crack growth rate of NR, SBR, NBR and BR were 0.9 x 10^{-5}, 6 x 10^{-5}, 1 x 10^{-5} and 50 x 10^{-5} cm/cycle respectively (Table 3). It was found that black-filled NR has the highest tear resistance as compared to other synthetic rubbers. Perhaps this is attributed by the large mechanical hysteresis in NR at high strains because of strain-induced crystallization (Lake, 1972). Hysteresis enhances the crack growth resistance of material by dissipating energy that would otherwise be expended in crack growth.

The crack growth rate dc/dN was found to be in the ranking order NR=0.9 x 10^{-5} > NBR=1 x 10^{-5}> SBR=6 x 10^{-5} > BR=50 x 10^{-5}. With the exception of strain-crystallized NR, the crack growth behavior of all amorphous rubbers in this study is highly influenced by the glass transition temperature. The glass transition temperature (Greensmith, 1956) T_g, is affected by the molecular mobility of the polymer which in turn reflects the internal viscosity of the polymer. It is well established that energy dissipation affects tensile and tear strengths. The higher the energy dissipated, the lower the energy available to cause crack propagation. Consequently, high input energy from external forces is required to compensate for energy dissipated. It is understood that the polymers with high glass transition temperature normally dissipate higher energy than those polymers with low T_g due to lower molecular mobility (high internal viscosity). For this reason, NBR which has the highest T_g shows the lowest crack growth rate dc/dN that is at 1 x 10^{-5} cm/cycle only. On the other hand, BR which has the lowest T_g in this case shows the highest crack growth rate at 50 x 10^{-5} cm/cycle or lowest crack resistance. In the case of NR, energy dissipation comes from strain-crystallization itself. This finding is in accord with Lake and Lindley (Lake et al., 1964b) where they have shown that NR was having the best crack growth resistance and BR was the worst.

Table 3. Crack growth rate (dc/dN) for different types of polymers.

Polymer	Tearing Energy (kJ/m^2)	dc/dN (cm/cycle)	Glass Transition Temperature, T_g (°C)[8]
NBR	0.85	1.00 × 10^{-5}	–40
BR	0.81	50.00 × 10^{-5}	–85
SBR	0.81	6.00 × 10^{-5}	–55
NR	0.88	0.90 × 10^{-5}	–68

Effect of Filler Loading

The effect of crack growth rates on various carbon black filler loading are shown in the Table 4. At lower filler loading that is 30 pphr, the dc/dN is at 14.00 x 10^{-5} cm/cycle. However, when the loading was increased to 60 pphr, the crack growth rate becomes very much smaller at only 2.00 x 10^{-5} cm/cycle (Table 4). Perhaps this can be explained the fact that, for a given particle size and filler structure, the degree of reinforcement is also affected by the quantity of filler incorporated into the rubber. The more reinforcement obtained, in general, the higher the hysteresis imparted because of increased opportunity for black-polymer bonds and/or the secondary network to be broken, which processes absorb energy (Lindley, 1973; Payne et al., 1971; Yoshihide Fukahori et al., 2003). Thus the higher the carbon black loading that is at 60 pphr, the more extensive is the secondary network and therefore the higher the hysteresis as compared to vulcanizate with 30 pphr loading. The increase in crack growth resistance with increasing amount of carbon black is attributed to blunting of the crack tip, or in an extreme case the crack tip forks out or splits into two or three crack tips. This result is consistent with the recent findings of Hamed, (Hamed et al., 2003) who used dumb-bell test piece to measure the growth resistance of NR vulcanizates.

Table 4. Crack growth rate (dc/dN) for different filler loading.

Black Loading	dc/dN (cm/cycle)
30	14.00×10^{-5}
40	10.00×10^{-5}
50	3.00×10^{-5}
60	2.00×10^{-5}

CONCLUSION

From the observations and results obtained, the following conclusions are drawn:

- The carbon black loading influences the crack growth resistance of NR vulcanizates that is better resistance at a higher loading.
- Strain-crystallization enhances crack growth resistance. In the absence of strain-crystallization, the crack growth resistance increases with the increasing order of the glass transition temperature, T_g.

KEYWORDS

- **Crack-growth cycle**
- **Rubber**
- **Split-tear test**
- **Vulcanizates**

ACKNOWLEDGMENT

The authors wish to acknowledge the Faculty of Applied Sciences, University Technology MARA Shah Alam and Continental Sime Tyre Technology Centre, Malaysia where some parts of the experimental works being carried in both organizations.

Chapter 7

Effect of Nanoparticles on Complexed Polymer Electrolytes

P. N. Gupta*, G. K. Prajapati, and R. Roshan

INTRODUCTION

Thin films of Nano-Composite Polymer Electrolyte (NCPE) consisting of poly (vinyl alcohol) (PVA), ortho phosphoric acid (H_3PO_4), and Al_2O_3 nanoparticles (size ~50 nm) have been prepared by solution cast technique. Solid Polymer Electrolytes (SPE) (PVA-H_3PO_4: 70:30 wt%) as first phase host matrix and nanoparticles of Al_2O_3 as second phase dispersed in different wt. ratios are utilized for the preparation of thin films of NCPE. The DSC result shows that both glass transition and melting temperatures of SPE films decreases with addition of Al_2O_3 nanoparticles. The FTIR spectra give indication about possible interactions between host matrix and dopants for its complexation. It has been found that ionic conductivity ($\sigma = 2.7 \times 10^{-4}$ S/cm) of the NCPE film having 2 wt% of nanoparticles is enhanced by one order of magnitude over the conventional SPE system at room temperature. The ionic transport behaviors in both SPE and NCPE films have been characterized and compared in terms of ionic conductivity (σ) and ionic transference number (t_{ion}) responsible for ionic conduction. The temperature dependent conductivity behavior has been used to compute the activation energy (E_a) involved in the conduction process.

Polymer electrolytes have drawn considerable attention worldwide due to their properties which make them a new class of promising materials suitable for large number of applications. In this regard Ion Conducting Polymer Electrolyte films play a significant role because of their light weight, easy processibility, high flexibility and relatively large ionic conductivity (Gray, 1991). Conventional SPE complexes dispersed with nano-sized filler particles are referred as nano composite polymer electrolytes (NCPEs). Dispersion of inorganic oxides such as Al_2O_3, SiO_2, TiO_2, and so on improves the mechanical as well as electrochemical properties of polymer electrolytes (Chen et al., 2000; Sharma et al., 2007). Nanocrystals leads to variety of exciting phenomena due to enhanced surface to volume ratio and reduced scale of transport lengths for both mass and charge transport (Appetecchi et al., 1995; Bohnke et al., 1992; Scrosati, 1993; Song et al., 1999). At elevated temperatures, ionic conductivity of the NCPE materials is increased substantially. This is generally attributed to the fact that nano fillers prevent the polymer chain reorganization and thus maintains the high degree disorder in the system (Agrawal, 2008). As a result much attention has

*Correspondence to: Prof. P. N. Gupta, Tel: +91-542-2307308, 2368390, Fax: +91-542-2368174, Email: gu02ptapn07@yahoo.co.in

been given on polymer electrolytes in general and amorphous polymer electrolytes in particular having low glass transition temperature (Christie et al., 2005).

In the present communication effect of nanosized filler particles of Al_2O_3 on electrical and dielectric properties of PVA based SPE have been investigated.

EXPERIMENTAL

The PVA (Mw ~ 1, 25,000, Aldrich), Orthophosphoric acid (H_3PO_4) (Ranbaxy, India) and nano particles of Al_2O_3 (50 nm, Sigma) were used for the preparation of NCPE films. The films of Solid Polymer Electrolyte (PVA + H_3PO_4) and Nano Composite Polymer Electrolyte (PVA + H_3PO_4 + Al_2O_3) in different filler wt% ratios were prepared by solution cast technique. Granules of PVA were dissolved in triply distilled water and stirred magnetically for 10–12 hr to obtain a clear and homogeneous solution. A known amount of H_3PO_4 was added to it and stirred continuously until a viscous solution was obtained. A part of solution was poured into the Teflon Petri dishes to get SPE thin film and in another part of the solution, nano-sized Al_2O_3 was added in different wt % to get NCPE thin films. The prepared thin films were rinsed with benzene/methanol to remove any volatile impurities present in it.

To know the possible complexation of Al_2O_3 nanoparticles with SPEs, films were characterized by FTIR and DSC spectra. The FTIR spectra were recorded for 400–4,000 wavenumber using Varian 3,100 FTIR Spectrometer (Excaliber series, German make) and DSC thermograms were carried out on Mettler Star SW 9.0 Differential Scanning Calorimeter having a nitrogen oven chamber over the temperature range +40 to 300°C. Electrical and dielectric properties were obtained using HP 4277 LCZ meter with a CT-806T temperature controller having accuracy of ±1°C.

RESULTS AND DISCUSSION

FTIR Studies

Figure 1 shows the FTIR spectra of SPE and NCPE having 1 and 2 wt% of nanoparticles. The spectrogram clearly indicates that the position of the transmission peaks for nano Al_2O_3 doped polymer electrolyte sample does not deviate to a large extent from that of un-doped electrolyte sample. The O-H stretching in SPE is observed at 3,438 cm^{-1} (Prajapati et al., 2010) while that in NCPE for 1 and 2 % Al_2O_3, the corresponding peaks are the same at 3,434 cm^{-1}. For better understanding of the IR spectra, peaks corresponding to C-H stretching, C-H bending and O-H bending have been assigned for all these three samples (Dyer, 1991) and listed in Table 1.

Table clearly indicates that formation of new peaks is not observed while the intensity of corresponding peaks decreases. It signifies that scissioning of polymeric bonds have taken place. Owing to chain scissioning, the mobility of polymer chains increases which in turn reduces the glass transition temperature. This trend shows that the doping of nano Al_2O_3 has not much significant contribution in altering the structure of host polymer matrix at molecular level. The absence of peaks corresponding to phosphate ions (PO_4^{3-}, HPO_4^{2-}, etc.) shows the complexation of PVA with dopant. This, in turn, suggests the enhancement in the concentration of mobile ions which leads to higher ionic conductivity in NCPE samples rather than SPE films. The spectra clearly indicate that peaks are present in their original position having small change in their intensity. This indicates that addition of filler only affects the morphology of electrolytes. Thus, nano Al_2O_3 particles are not complexed with host polymer but only support the movement of mobile species through the polymer backbone. This can be easily seen from conductivity variations also.

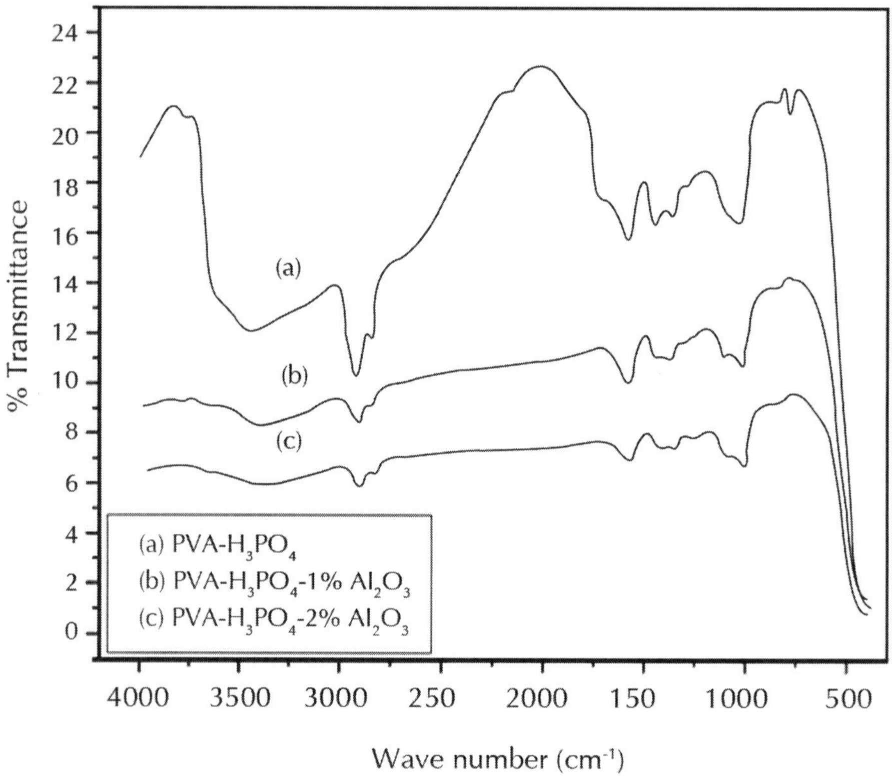

Figure 1. FTIR spectra of various polymer electrolyte samples.

Table 1. FTIR band assignment for SPE and NCPE films.

Band assignment	$PVA-H_3PO_4$	$PVA-H_3PO_4$+ 1% Al_2O_3	$PVA-H_3PO_4$+ 2% Al_2O_3
O-H stretching	3438	3434	3434
C-H stretching	2924	2923	2923
C-H & O-H bending	1381	1382	1383
O-H bending	1038	1037	1036

DSC Studies

Figure 2 shows the Differential Scanning Calorimetry curves for SPE and NCPE thin films. Figure reveals that addition of Al_2O_3 nanoparticles to the SPE, glass transition temperature (T_g) of NCPE decreases. The observed decrease in T_g for both types of the sample is due to increase in amorphicity as well as free volume of the sample with the addition of filler (Kang et al., 2006). It is found that the sample containing 1% nano Al_2O_3 have three melting peaks at 158.89, 162 and 178.92°C whereas sample containing 2% Al_2O_3 shows only two peaks at 137.16 and 153.25°C. The appearance of more than one melting peak may be due to different crystalline regions, extent of crystallinity and nature of crystalline morphology present in the electrolyte sample (Pandey et al., 2008). Moreover, there is significant decrease in the melting temperature with increase in Al_2O_3 concentration. This is quite expected as Al_2O_3 reduces the force between the polymer chains as it remains interdispersed between the chains. This, in turn, leads to increase in conductivity of the sample due to increase in their free volume for the conduction of mobile species, as depicted from conductivity variations.

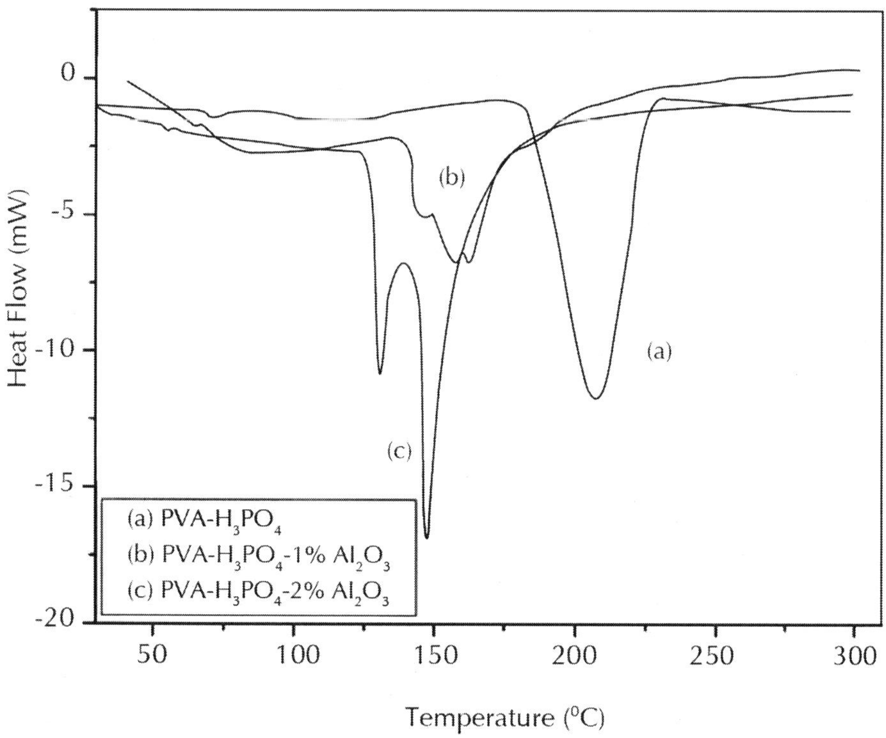

Figure 2. DSC thermograms of various polymer electrolyte samples.

Transference Number Measurement

The ionic transference number of SPE and NCPE thin films has been estimated by Wagner's polarization technique (Wagner et al., 1957) for which the likely mobile species are protonic. In the polarization technique, a potentiostatic current is recorded as a function of time domain across the cell configuration electrode/sample/electrode and shown in Figure 3. The transference number of ions (t_{ion}) has been calculated using the relation

$$t_{ion} = \frac{I_T - I_e}{I_T} = \frac{I_{ion}}{I_T} \quad (1)$$

where $I_T = I_e + I_{ion}$ is the total current (sum of electronic and ionic currents) and I_e is the electronic current. Using the above relation, the ionic transference number for both electrolyte samples has been calculated and is listed in Table 2. Table shows that t_{ion} for both SPE and NCPEs approaches to unity which strongly confirms their ionic (protonic) nature (Prajapati et al., 2010).

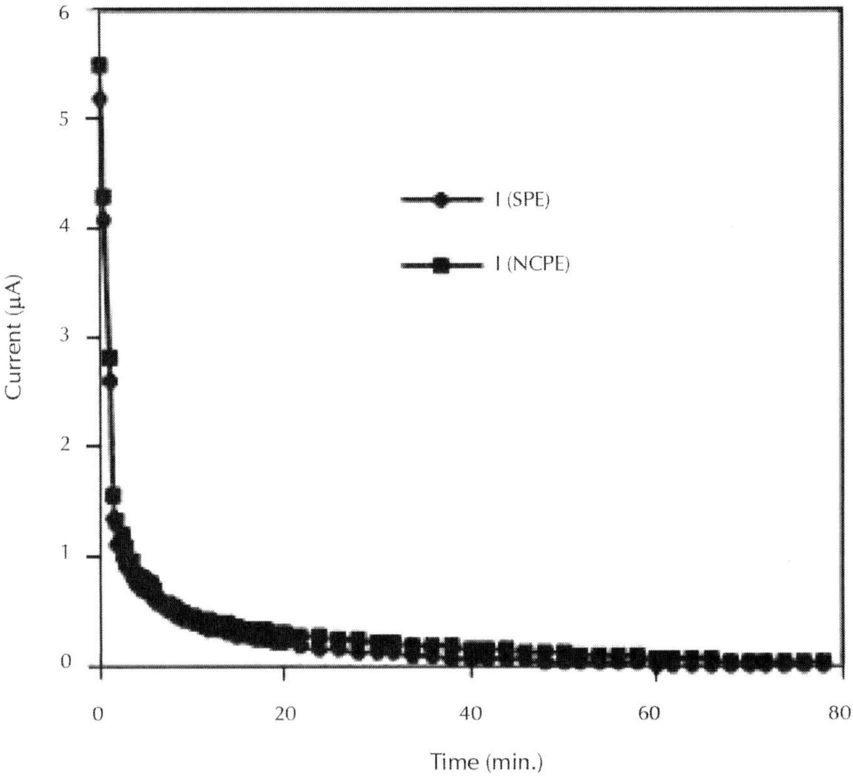

Figure 3. Polarization current as a function of time for SPE and NCPE thin films.

Table 2. Transference number and activation energy for NCPE thin films.

NCPE films	Transference number (t_{ion})	E_a (eV)	
		Below T_g	Above T_g
0% nano Al_2O_3	0.92	0.33	0.28
1 % nano Al_2O_3	0.95	0.30	0.26
2 % nano Al_2O_3	0.96	0.29	0.22

Ionic Conductivity Analysis

The bulk conductance and therefore bulk conductivity of SPE and NCPE thin films has been calculated using complex admittance (*B-G*) spectroscopy technique (Chandra et al., 1983). This technique is an effective tool for monitoring interfacial phenomena. A typical *B-G* plot for NCPE film is shown in Figure 4. Using *B-G* plot, the bulk conductivity (σ) has been calculated from the relation

$$\sigma = G\frac{d}{A} \tag{2}$$

where d and A are the thickness and cross sectional area of the sample respectively. Figure 5 shows the temperature dependent conductivity plot for SPE and NCPE thin films over the temperature range 303–393 K. The plot shows two distinct straight lines of Arrhenius behavior separated by a slightly curved region in the temperature range 343–353 K. Similar characteristics behavior has also been reported by other workers (Bhargav et al., 2007; Singh et al., 2004). The sharp change in conductivity started around 70°C may due to transition from crystalline to amorphous phase where polymer goes into visco-elastic phase (Pandey et al., 2008). Consequently, the curved regions in the plot are caused due to glass transition of the sample from hard glassy state to more amorphous rubbery state. Therefore, as temperature increases, free volume for the motion of hydrogen ions through the polymer backbone also increases (Sekhon et al., 1998) and hence ions can easily jump into its neighboring vacant sites

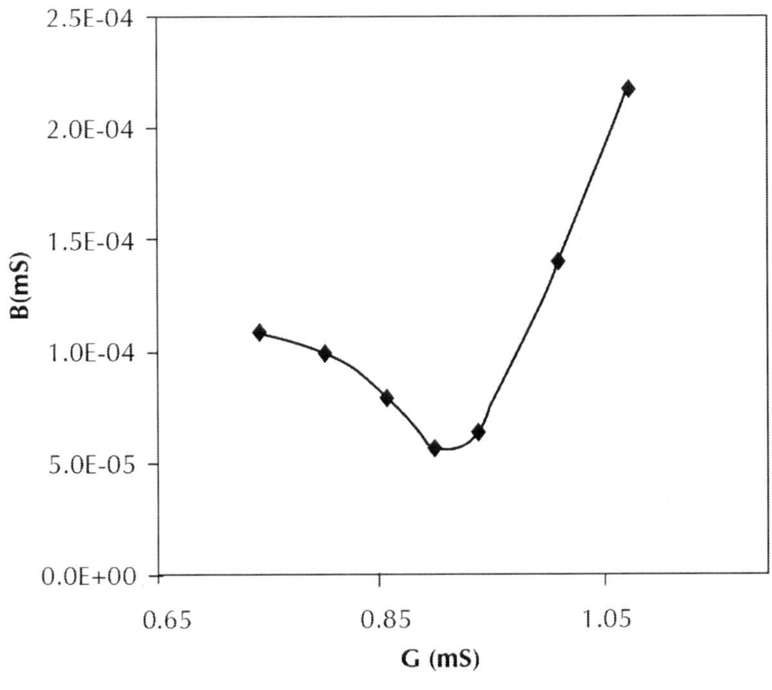

Figure 4. Typical B-G Plot for 2wt% of NCPE film.

(Howe et al., 1979). As a result, the ions and polymer segments can move faster in the free volume which enhances the ionic conductivity as the temperature increases (Souquet et al., 1994). Moreover, the linear variation in the characteristics curve before and after the glass transition temperature suggests an Arrhenius-type thermally activated process. Figure 5 also shows that conductivity increases due to addition of nano fillers and show evidence of maximum conductivity (2.7×10^{-4} S/cm) for 2 wt % of nano filler Al_2O_3. The increase in ionic conductivity with addition of nano fillers may be due to the increase in amorphicity of the electrolyte samples (Ahmad et al., 2009). This may be also inferred from the minimum activation energy for 2 wt% NCPE sample as presented in Table 2. It is found that activation energy is low in amorphous region as compared to semicrystalline region. Such features are not observed in inorganic larger size filler particles.

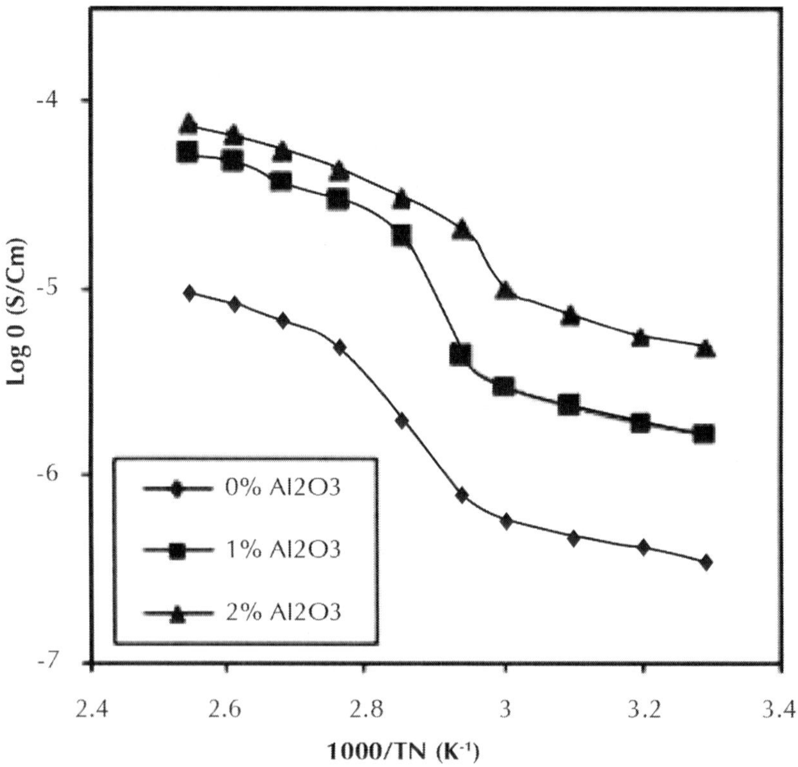

Figure 5. Temperature dependence of bulk ionic conductivity for SPE and NCPE thin films.

Dielectric Constant Studies

The dielectric constant (ε') of the thin films has been calculated using the relation

$$\varepsilon' = \frac{C.d}{\varepsilon_0 . A} \tag{3}$$

where ε_o is the permittivity of free space and C is the capacitance measured from LCZ meter. The variation of ε' with temperature at two constant frequencies for both type of electrolyte samples (SPE and NCPE) is shown in Figure 6. It is found that dielectric constant increases with temperature at lower rate in the high temperature range for a particular frequency. This can be attributed due to decrease in viscosity of the electrolyte sample with temperature that leads to enhancement of the dipolar rotation (Mohamad et al., 2006). It is also observed that the value of ε' is quite large for NCPE sample in comparison to SPE films. This can be linked to an enhancement in amorphous nature of the electrolyte sample. The amorphous phase of polymeric material facilitates the orientation of the dipoles present in the electrolytes and causes high value of dielectric constant.

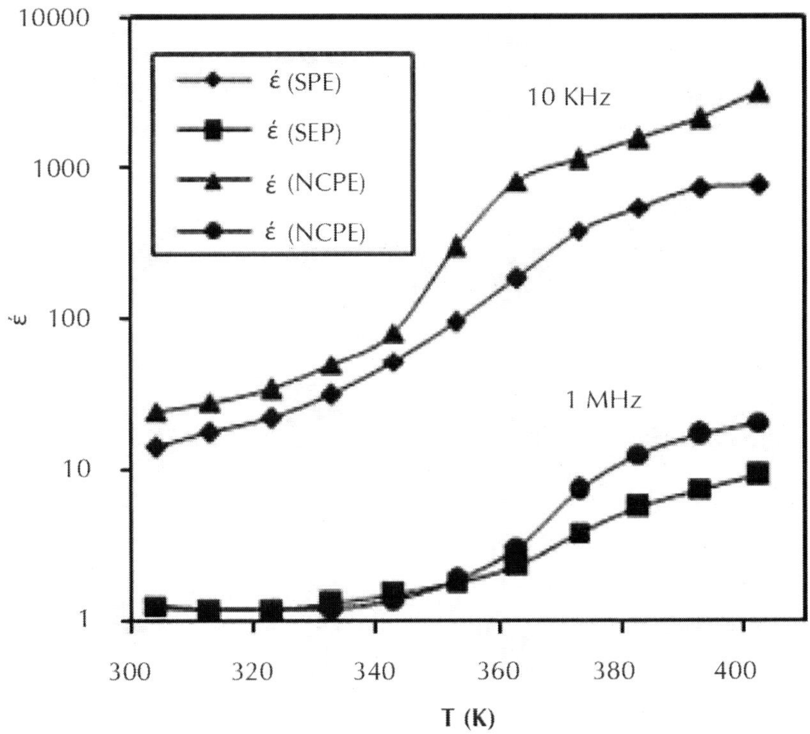

Figure 6. Variation of dielectric constant vs. temperature for SPE and NCPE thin films.

Figure 7 shows the variation of dielectric constant with frequency at room temperature for SPE and NCPE films. It is observed that dielectric constant decrease with increasing field frequency. High value of dielectric constant at low frequency can be explained by the presence of space charge effects due to accumulation of enhanced charge carrier density near the electrode. As frequency increases, the periodic reversal of the electric field occurs so fast that there is restriction of excess ion diffusion in the direction of applied field and hence dielectric constant decreases. Thus conductivity increases to a large extent by the addition of nano particles. Figure also reveals that dielectric constant first decreases very rapidly in low frequency range while it decreases slowly in the range of high frequency (Akaram et al., 2005; Singh et al., 1998).

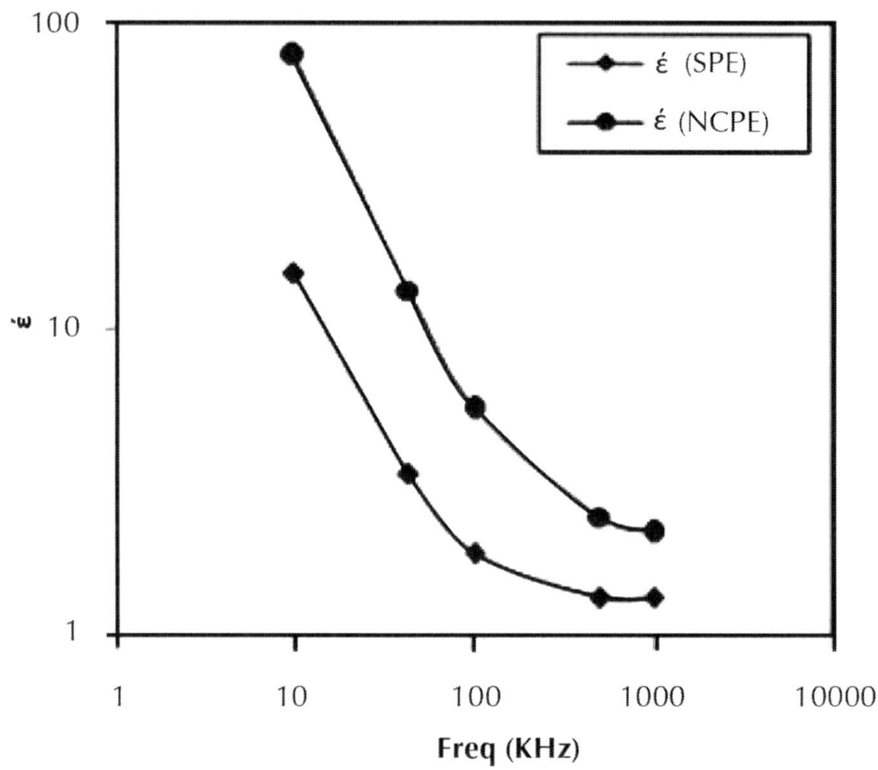

Figure 7. Variation of dielectric constant with frequency for SPE and NCPE films.

CONCLUSION

Effect of nano particles of Al_2O_3 on conventional SPE films have been examined by FTIR, DSC and B-G spectroscopy. The dispersal of Al_2O_3 nano particles to the SPEs shows declination in the glass transition and melting temperature as established from DSC analysis. The FTIR spectra show possible interactions between Al_2O_3 nano particles and host SPE films. The optimum room temperature ionic conductivity of the order of 7×10^{-4} S/cm having minimum activation energy ($E_a \sim 0.22$eV) is observed for NCPE films. This shows one order increment in the conductivity over the conventional SPE films. The temperature dependent conductivity shows Arrhenius type thermally activated behavior before as well as after glass transition temperature. Maximum value of ion transference number is found to be 0.96 which is indicative of predominant ionic (protonic) transport in the SPE and NCPE thin films. It has been observed that dielectric constant for SPE and NCPEs increases with temperature while it decreases with frequency.

KEYWORDS

- **Activation energy**
- **Electrical conductivity**
- **Nano composite polymer electrolytes**

ACKNOWLEDGMENT

One of the authors GKP is thankful to CSIR, New Delhi for award of Senior Research Fellowship.

Chapter 8

Control the Flow Curve of the Thermoplastic Vulcanized (PE/PB)

Moayad N. Khalaf, Hussain R. Hassan, Raed K. Zidan, and Ali H. Al-Mowali

INTRODUCTION

Thermoplastic vulcanized polybutadiene/high density polyethylene (TPV/HDPE) (70/30) was prepared using organic peroxide (3%) as vulcanizing agent. A maleated polyethylene (MAPE) was prepared by melt mixing. The percent of maleation found was (1.78%). The percent of MAPE added to the (TPV) was (0, 2, 4, 6, 8, and 10). The rheological properties were measured at temperature of 140°C. Discontinuity of the shear stress vs. shear rate curve appeared for 0, 2, 4, and 6% of the MAPE. While for 8% of the MAPE the curve was continuous with concavity at the second critical shear stress (τ_2) and a complete elimination of curve discontinuity was reached when using 10% MAPE.

Melt flow instabilities, loosely referred to as "melt fracture", are phenomena limiting polymer extrusion. The general trend in the development of flow instabilities during extrusion of polymer melts involves the absence of extrudate surface distortions at relatively low shear rate/stress values (at which presumably the wall slip velocity is negligible) on one hand and at relatively very high shear rate/stress values at which correspondingly high (and presumably stable) wall slip velocities exist. This suggests that extrudate distortions are less likely to occur under conditions in which a stable flow boundary condition exists at the wall; whether it is either a stable wall stick or stable wall slip condition. The occurrence of these instabilities for plastics is known at least since 1945 (Rudolf, 1998). The "sharkskin effect" in polymer extrusion was the first instability to occur and thus it is the first problem that must be solved, before one needs to worry about the other instabilities. The manufacturing of plastics is one of the most important industrial manufacturing processes, and is used to create a huge variety of objects (rods, wire, sheet, fiber, and others). The most common process is extrusion, in which polymer fibers or sheets are produced by forcing the polymer melt or solution through a cylindrical hole or slit (a"die") (Denn, 2001). Flow instabilities that occur in melt processing arise from a combination of polymer viscoelasticity and the large stresses that occur and cause large, rapid deformations. This is in contrast to flows of Newtonian fluids, where inertia and surface tension is usually the driving forces for flow instabilities. Both the elasticity and the high stresses that occur in the flow of molten polymers arise from their high molecular weight (i.e., from the enormous length of their molecules). The high stresses are associated with the high viscosity of molten commercial thermoplastics and elastomers. Typical values range from 10^3 to 10^6 Pa s, whereas the viscosity of water is about 10^{-3} Pa s. An easily observed manifestation of melt elasticity is the large swell in cross section that occurs when a melt exits a die.There are several common themes that emerge in the various extrusion instabilities and consequent techniques for their elimination. For example, wall slip plays a prominent role in sharkskin, stick–slip, and in the use of additives to provide stable flow. Processing aids (PAs) are frequently used. The PAs eliminate flow instabilities or postpone them to higher flow rates. The end result is an increase of the productivity as well as an energy cost reduction, while high product quality is maintains edge (Evdokia et al., 2002). The melt fracture of LLDPE was studied by (Denn, 1990, 2001) and found that the surface of the extrudate becomes visibly rough at a wall shear stress level that is typically of the order of 0.1 MPa, with apparent periodicity in the small-amplitude distortion; this phenomenon is commonly called sharkskin. At a higher level of stress, typically of the order of 0.3 MPa, the flow becomes unsteady and the extrudate alternates between sharkskinned and smooth segments; this is commonly called stick-slip or spurt flow. At still higher stress levels, sometimes after a second region of spurt flow, the flow becomes steady. The extrudate surface is relatively smooth during the early part of this steady regime, with a long-wavelength distortion, but gross distortions occur at higher stresses; this regime is commonly called wavy or gross melt fracture.

EXPERIMENTAL

Materials

High density polyethylene (HDPE) SCPILEX 5502 and Low Density Polyethylene (LDPE) SCPILEN 464 were supplied by state company for petrochemical industry (SCPI) in Basra (MFI = 0.46 gm/10 min, density = 0.953 gm/cm^3 and MFI = 4.0 gm/10 min, density = 0.9220 gm/cm^3 respectively). Polybutadiene was obtained from Malaysian company with (Mooney viscosity ML (1 + 4) at 100°C 45 ± 5). Maleic anhydride (MA) industrial grade was supplied by Adrash Chemicals and Fertilizers Ltd., Udhna Gujorat-India. The MA was used without purification. The 2-ethyl terbutyl-hexanoate peroxide (C67), peroxide content 98% was supplied by Pergan GMBH Company.

Instrument

Rheological Measurements

Rheological properties were carried out by using a capillary rheometer device (Instron model 3211), ASTM D-3835. The diameter of the capillary is 0.76 mm, the length to diameter (L/D) ratio of 80.9, with an angle of entry of 90°. Load weighing which dropped on the polymer melts by plunger transversed from the top to the bottom of the barrel was constant (2,000 kg). The constant plunger speeds ranged from 0.06 to 20.0 cm.min^{-1}, the extrusion temperature was 140°C.

Mixer Instrument

Mixer-600 attached to Haake Rhechard Torque Rheometer supplied by Haake Company. The RPM of the Mixer can be controlled depending on the shear rate.

FTIR-Measurements

Fourier transforms infrared spectrophotometer Shimadzu FTIR-84005 was used for measuring the IR-Spectra for the samples.

Polyethylene (PE)-Maleic Anhydried Grafting

The LDPE Grade 464 was fed to a Mixer-600 attached to Haake Record meter at temperature = 150°C and RPM = 64. After 5 min, MA was fed portion wise and the peroxide was added drop wise. The modified PE was characterized by FTIR and evaluated via the degree of maleation (%) according to (Ghaemy et al., 2003; Moayad et al., 2008).

$$\text{Acid Number (mg KOH/g polymer)} = \frac{\text{ml KOH x N KOH x 56.1}}{\text{polymer(g)}}$$

Therefore

$$\%\text{MAH} = \frac{\text{Acid no.X98}}{\text{2X56.1}}$$

MAH=1.78%

Preparation of the Thermoplastic Vulcanized (PE/PB)

Polybutadiene was feed to Mixer-600 at temp = 140°C and RPM = 32. The HDPE and the MAPE mixed together before added to the Mixer-600, after 2 min the peroxide was added. Then the velocity of mixing was changed to RPM = 64 and the mixing time continue for 10 min.

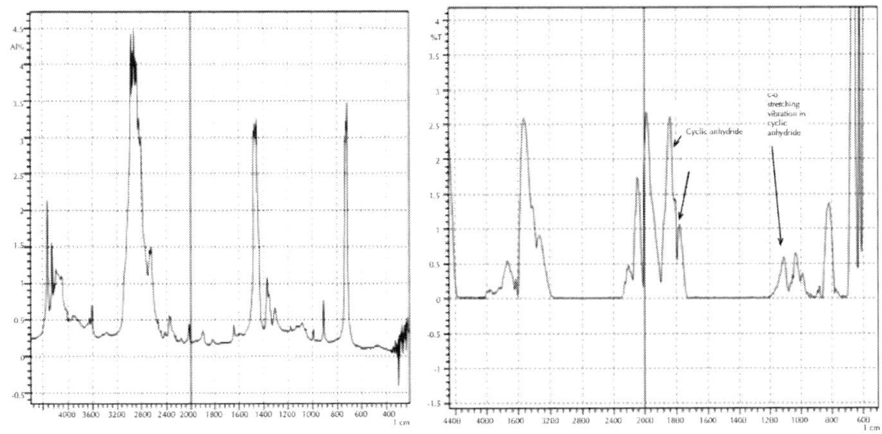

Figure 1. FT-IR spectra of (A) polyethylene and (B) Maleated polyethylene.

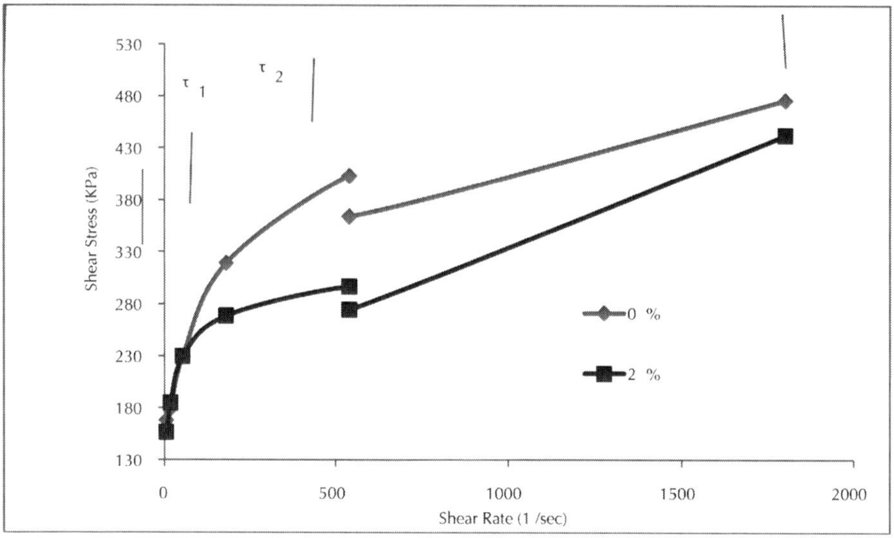

Figure 2. Flow curve of TPV at 140°C with 0% (MAPE) and with 2% (MAPE).

RESULT AND DISCUSSION

The comparison of the infra-red spectra of the pure PE and the MAPE sheet, Figure (1 (a), and (b)) has confirmed the presence of MA units on PE backbones. In comparison with the infra-red spectrum of pure PE, the grafted PE showed very strong bonds at 1,779 cm^{-1} and 1,859 cm^{-1}, which are characteristics of cyclic anhydride. The band at 1,221 cm^{-1} corresponds to C-O stretching vibrations in the cyclic anhydride (Ghaemy et al., 2003; Moayad et al., 2008).The Flow curve instability phenomena for the TPV without compatabilizer agent (MAPE) was shown in Figure 2. From Figure 2 the flow curve was divided into four regions, the first region started at shear stress (168–179) KPa and shear rate (5.4–18 S^{-1}) the curve was smooth, the second region started at stress (229.4–319.2) KPa which was the first critical shear stress(τ_1) and shear rate (54–180 S^{-1}), there was fluctuation in the curve and sharkskin phenomena was observed (Humberto et al., 2009, 2010; Moayad et al., 2008). The third region started at shear stress (364–403.2) KPa the second critical shear stress(τ_2) and shear rate (540 S^{-1}) there was pressure oscillation and stick-slip phenomena was observed (Denn, 2001),while the fourth region started at shear stress (476) KPa and shear rate (1,800 S^{-1}) and melt fracture phenomena was observed. The fluctuation in flow curve due to the uncompatability between the polymer blend and adding 2% of the MAPE as compatabilizer the curve was linear Figure 2 and the value of the shear stress was reduce to (235.2–285.6) KPa, this behavior was attributed to the increase of the chain interactions between the two polymers. While the second critical shear stress (τ_2) at shear rate 540 S^{-1} the value was reduced to (274.4–296.8) KPa, but the rheological distortion still appeared because it was based on a stick-slip mechanism. It is purported (Jaewhan Kim et al., 2009; Lau et al., 1986; Ramamurthy, 1986) that, above a critical shear stress, the polymer experiences intermittent slipping due to a lack of adhesion between polymer and die wall, in order to relieve the excessive deformation energy adsorbed during the flow. There was pressure oscillation inside the capillary and discontinuity was appeared on the flow curve for the polymers blend with (4 and 6%) compatabilizer (MAPE) as shown in Figure 3. While increasing the percent of the MAPE to 8% curve was linear and smooth as shown in Figure 4.The 8% was the sufficient quantity to provide the sufficient chains interaction between the two polymers blend and not the sufficient adhesion force between the polymers blend and die surface which causes appeared of concavity on the flow curve. While the 10% compatabilizer (MAPE) was the sufficient quantity to achieve the chain interaction between the two polymer and the force adhesion between the polymers blend and the die surface. Result in the disappearance of flow instability curve of polymers blend. The increase in adhesion force is due to the presence of the carboxyl group and hydroxyl group of the MA on the backbone of the PE, thus improved the adhesion property between the TPV and the die Figure 5.

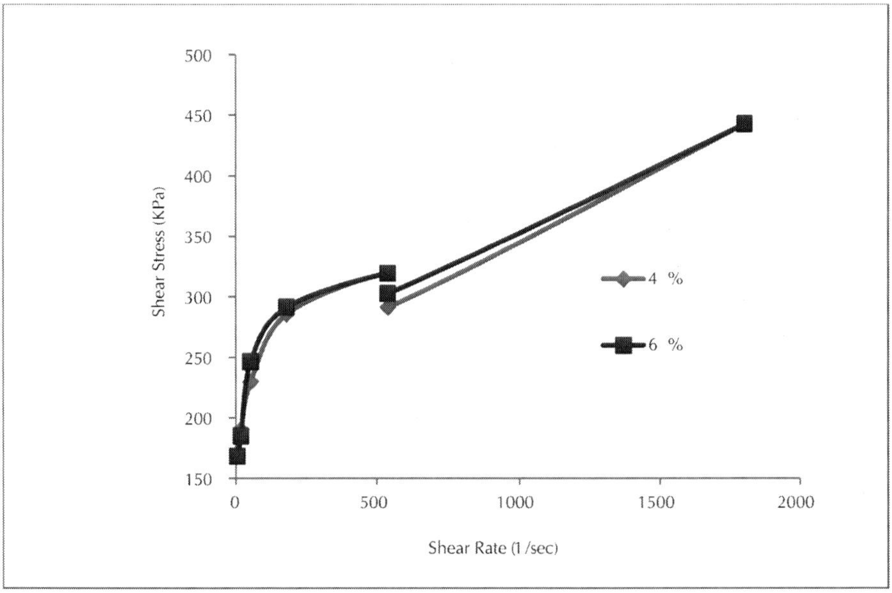

Figure 3. Flow curve of TPV at 140°C with 4% (MAPE) and with 6% (MAPE).

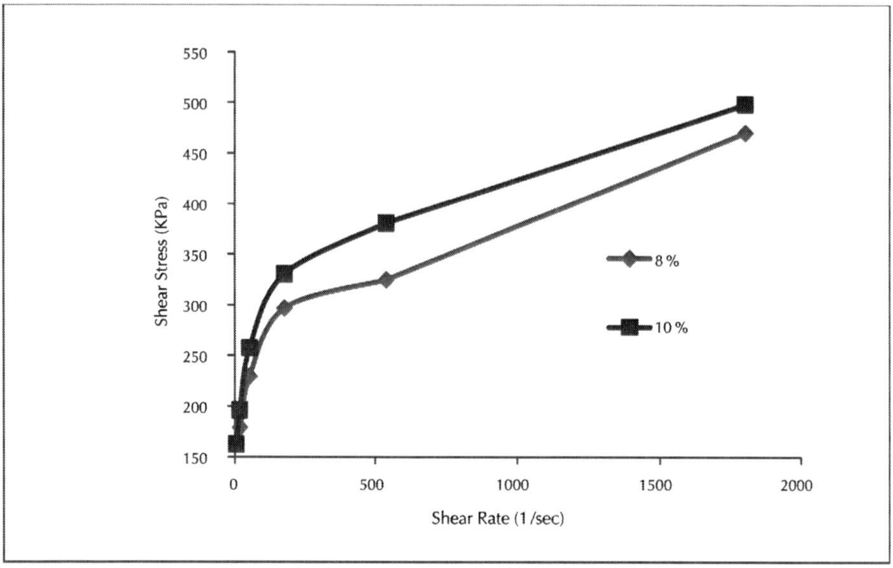

Figure 4. Flow curve of TPV at 140°C with 8% (MAPE) and with 10% (MAPE).

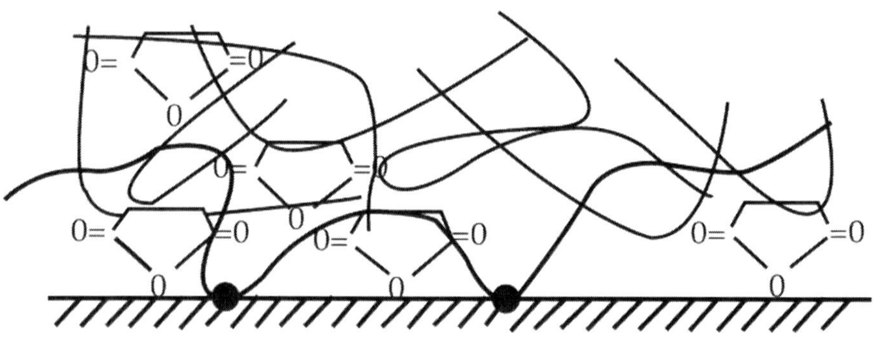

Figure 5. Schematic representation of maleic anhydride units on polyethylene backbones, which increases the adhesion force between the TPV blend and die surface.

CONCLUSION

The effect of compatabilizer (MAPE) to eliminate discontinuity in the flow curve of the thermoplastic vulcanized (PE/PB) was studied. Various concentration of the compatabilizer (0, 2, 4, 6, 8 and 10 wt/wt%) were tested. It was shown that the complete elimination of discontinuity which caused by the pressure oscillation inside the die during the polymer melt flow. The presence of the MAPE will act as compatabilizer between the two polymer chain and both the 8% and 10% can achieved the complete chain interaction and the blend will be completely miscible and act as one phase, while the 8% cannot offer the sufficient adhesion forces between the polymer and the die wall, for that there was concavity in the flow curve, which disappeared in the flow curve of TPV with 10% MAPE.

KEYWORDS

- **Polymer melt instabilities**
- **Polymer rheology**
- **Thermoplastic vulcanized**

Chapter 9

Solid-state Scatty Lasers Based on Nanohybrid POSS Composites

A. Costela, I. García-Moreno, L. Cerdán, V. Martín, M. E. Pérez-Ojeda, O. García, and R. Sastre

INTRODUCTION

Progress in the chemical synthesis of hybrid matrices has allowed the design and construction of new optical and optoelectronic materials with specific properties for applications in areas of high technical importance, such as light emitting diodes (Burroughes et al., 1990), field effect transistors (Garnier et al., 1994), photodetectors (Kanicky, 1986), solar cells (Huynh et al., 2002) or laser materials (Reisfeld, 2004). In the attempt to develop solid-state dye lasers (SSDL), highly emissive structures based on hybrid materials doped with laser dyes have been synthesized via the sol-gel technique (García et al., 2008; Reisfeld, 2004). In this way, it is possible to combine in a unique material the advantages of inorganic glasses (high thermal dissipation capability, low thermal expansion and thermal coefficient of refractive index *dn/dT*, high damage threshold) (Barnes, 1995; Nikogosian, 1997; Rahn and King, 1998) with those offered by organic polymers (high capability to solve organic dyes, good homogeneity, adaptability to techniques of cheap production, easiness to modify the material's composition and chemical structure) (Costela et al., 1996, 2001; Duarte, 1994; Gómez-Romero and Sánchez, 2004; Rahn and King, 1998; Sastre and Costela, 1995). However, these materials exhibit some serious limitations such as complex and lengthy synthesis process, fragility that makes mechanization and polishing of the final material difficult and, most important, frequent optical in homogeneity caused by refractive index mismatch between organic and inorganic domains.

A way to avoid these problems while maintaining the combining advantages of polymers and inorganic materials has been the synthesis of silicon-containing organic matrices (Costela et al., 2007). In this approach, organic monomers are used with silicon atoms directly incorporated into their structure. Although high lasing photostabilities were obtained with these dye-doped matrices, the lasing efficiencies were lower than those obtained with some hybrid materials. In addition, the benefits induced by the presence of silicon are limited because of the low silicon content allowed to avoid obtaining very soft final materials unfit for subsequent mechanization and proper polishing.

Enhancement of photostability of organic dye molecules has been observed in polymer matrices co-doped with dielectric oxide microparticles (Ahmad, 2007). Duarte and James (2003, 2004) demonstrated a class of dye-doped organic-inorganic solid-state gain media based on silica nanoparticles uniformly dispersed in poly(methyl methacrylate) (PMMA), which exhibited better thermal parameters and improved optical homogeneity than previous composite gain media. As a result, high laser conversion efficiency and low beam divergence were obtained. These are as follows approach, we reasoned that improvement in homogeneity, efficiency, and photostability could be attained by incorporating silica at the molecular level by using monomers based on Polyhedral Oligomeric Silsesquioxanes (POSS) functionalized with methacrylate groups in their surface. These compounds have attracted much attention over the last years as molecular silica and nano-scaled building blocks for preparing nanocomposites with novel applications in photonics and electronic devices (Cordes et al., 2010; Lickiss and Rataboul, 2008; Pielichowski et al., 2006). The presence of POSS enhances significantly the thermal, mechanical, and physical properties of the final materials, opening the possibility to synthesize new luminescent hybrid matrices with optoelectronic properties comparable to those of dendronized or grafted conjugated polymers (Markovic et al., 2008; Zhao et al., 2008).

The POSS have a compact hybrid structure with a well-defined cage-like inorganic core made up of silicon and oxygen $(SiO_{1,5})_n$ externally surrounded by non-reactive or reactive polymerizable organic ligands, R (Scheme 1). The type and number of substituents control the interactions between the organic ligand and the medium, defining the compatibility and thus the final properties of the POSS-modified systems. The POSS nanoparticles can be dispersed at molecular

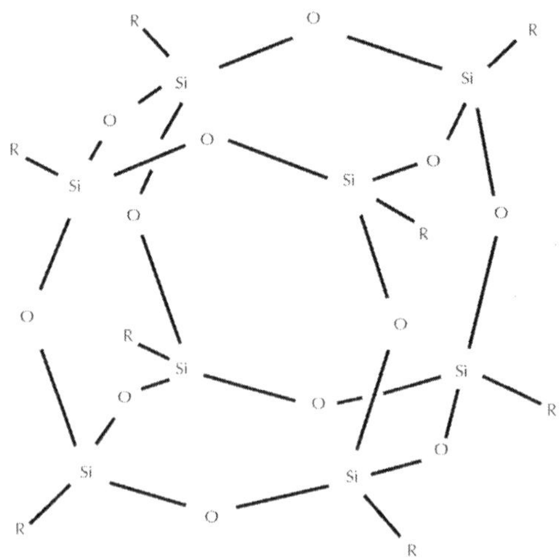

Scheme 1. Octameric POSS. level (1–3 nm) (Bizet et al,. 2006a) and, because of their synthetically well-controlled functionalization, can be incorporated into polymers by different polymerization techniques with minimal processing disruption (Kopesky et al., 2004; Zheng et al., 2002). This excellent dispersion at molecular scale and the copolymerization with organic monomers prevents phase separation, assuring the macroscopical homogeneity of the materials (Bizet et al., 2006b). In this way, new optical hybrid materials based on POSS nanoparticles as inorganic part overcome some of the most important limitations intrinsic to sol-gel hybrid composites while maintaining the physical, chemical, and mechanical combined advantages of organic-inorganic-systems (García et al., 2008).

Here, we present results showing remarkable improvements in the laser action of both polar and non-polar organic dyes in liquid and solid solutions incorporating POSS nanoparticles. Evidence will be presented that, under certain conditions, POSS nanoparticles added to solid and liquid solutions of organic dyes can scatter light in a way that increases the efficiency of the laser action of the material. The POSS particles act as weak scattering centers in the Rayleigh limit (particle size much smaller than the emission wavelength), and the photon path enlargement caused by multiple scattering provides an extra feedback which enhances incoherently the magnitude of the amplification process, in what has been called "Non-Resonant Feedback Lasing" (NRFL), "Incoherent Random Lasing", or "Lasing with Intensity Feedback" (Noginov, 2005; Takeda and Obara, 2009). A theoretical model for laser systems with weak scattering centers is developed. The model provides inklings to understand the physics responsible for the enhanced emission of these rather complex systems, and reveals that the key factor in the process is the nanometer size of the scatterers.

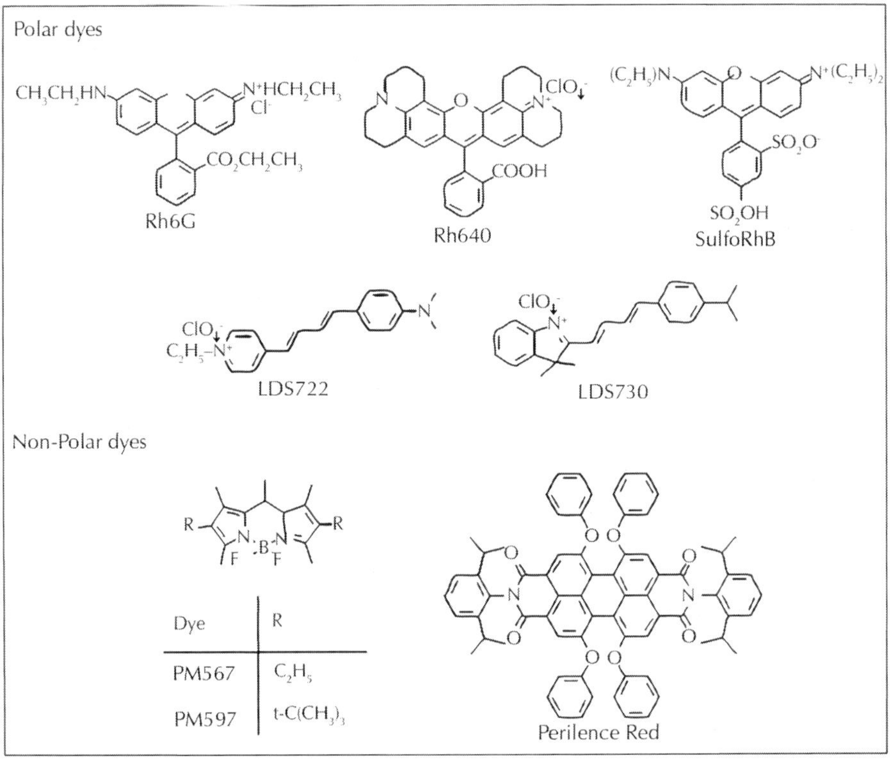

Figure 1. Molecular structures of laser dyes selected in the present work.

When the dye-doped gain media are deposited in the form of thin films onto glass substrates, defining a leaky waveguide structure, the presence of POSS allows obtaining laser emission without the need of incorporating complex resonant substructures in the material.

EXPERIMENTAL

Dyes of the pyrromethene, rhodamine, perylene, and styryl families (Figure 1) were incorporated into solid matrices or dissolved in appropriated solvents at concentrations in the 10^{-2} to 10^{-3} M range. Bulk and thin film solid samples consisted of dye-doped copolymers of methyl methacrylate (MMA) with octa(propyl methacryl)-POSS (8MMAPOSS) in weight proportions (wt%) from 1 to 50%, which corresponds to weight content of silicon from 0.2 up to 8% (Figure 2). The nomenclature used to name the materials is as follows: COP(MMA-8MMAPOSS *x*:*y*), where *x* and *y* are the weight proportions of MMA and 8MMAPOSS in the matrix.The synthesis route followed to prepare the materials is described in (García et al., 2008). Details of the preparation of liquid solutions, bulk samples and thin films for laser studies can be found in (Costela et al., 2009) (supporting information) and (Cerdán et al., 2009). Description of the experimental set-up used for laser evaluation can be found in (Costela et al., 2009) (supporting information) and (Costela et al., 2003) for liquid solutions and solid bulk samples, and (Cerdán et al., 2009) for thin films.

RESULTS AND DISCUSSION

A detailed characterization of the copolymers of MMA with different weight proportions of 8MMAPOSS was carried out in (Sastre et al., 2009). The density of the final material, the refractive index, and the thermal conductivity were found to increase linearly with the 8MMAPOSS content (Figure 3).

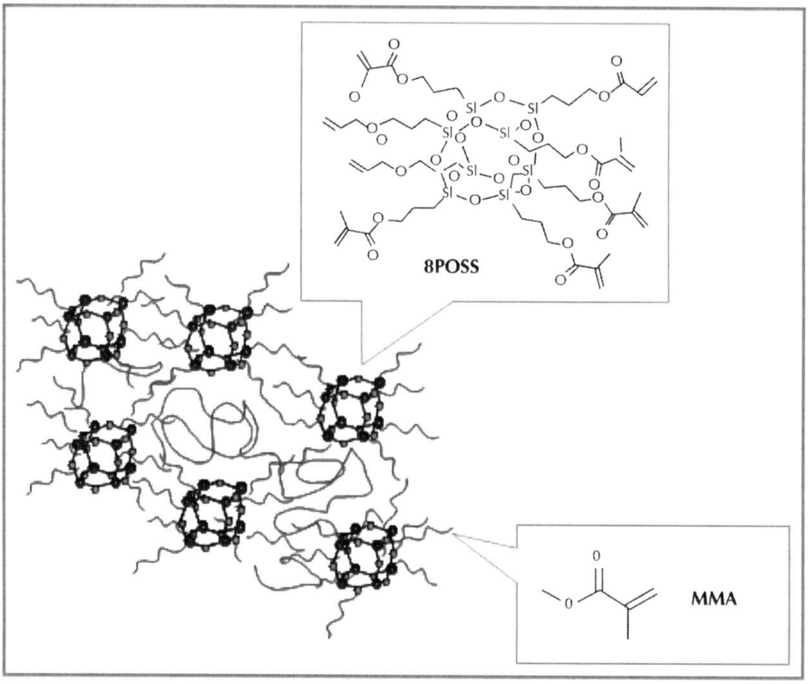

Figure 2. Schematic representation of MMA-8MMAPOSS cross-linked hybrid materials.

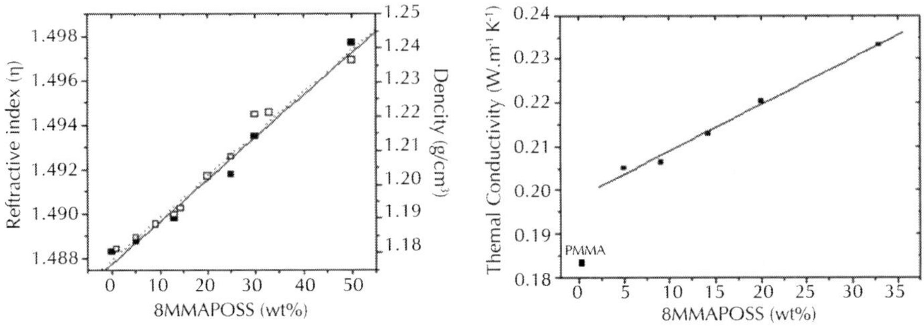

Figure 3. Variation of density (left: white squares and dashed line), refractive index (left: black squares and solid line), and thermal conductivity (right) of COP (MMA-8MMAPOSS) matrices as a function of 8MMAPOSS content.

The size of the 8MMAPOSS particles was in the nanometer/few nanometers range and they were homogeneously distributed. In Figure 4 is presented, as an example, the result of a dynamic light scattering (DLS) analysis of the size distribution by volume of 8MMAPOSS particles incorporated to a MMA solution in 13 wt % proportion. It can be appreciated that the 8MMAPOSS size distribution was centered at 1.17 nm, with an approximate range from 0.5 to 4 nm

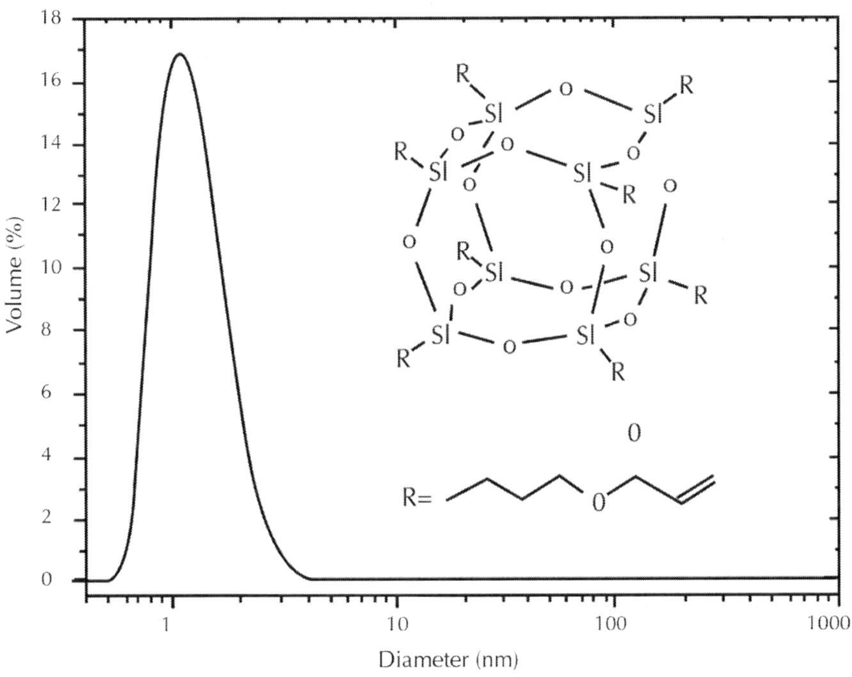

Figure 4. Size distribution of 8MMAPOSS particles incorporated in a 13 wt % proportion to a MMA solution.

Transmission electron microscopy (TEM), atomic force microscopy (AFM), and scanning electron microscopy (SEM) images of thin film sections of the gain media used the solid-state laser experiments revealed a homogeneous distribution of 8MMA-POSS in the polymer, as illustrated in Figure 5 for a sample containing 13 wt % of 8MMAPOSS, and depict a continuous phase that corresponds to the organic matrix which incorporates the nanosized POSS crosslinkers at a molecular level, which is in agreement with the transparency of the sample.

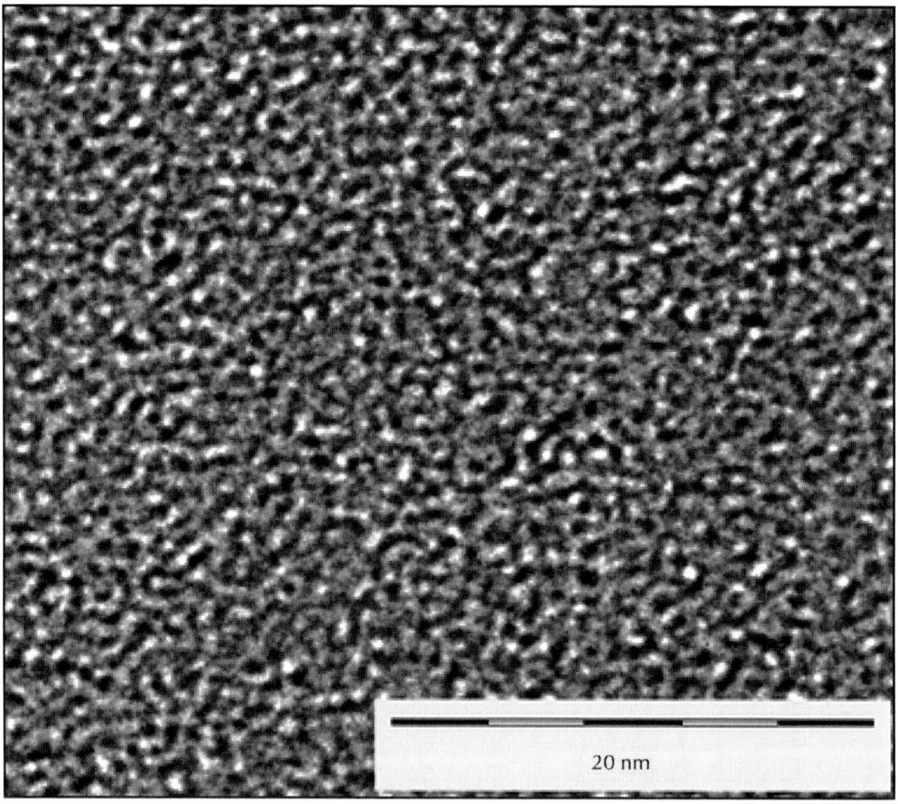

Figure 5. The TEM image of the COP (MMA-8MMAPOSS 87:13) network.

Laser Emission from Gain Media Incorporating POSS Nanoparticle

The presence of POSS nanoparticles was found to enhance the laser action of dyes of very different families both polar (Rhodamine 6G, Rhodamine 640, Sulforhodamine B, LDS722, LDS730) and non-polar (Perilene Red, PM567, PM597) with respect to the effectiveness recorded in pure organic solvents and polymeric solutions pumped under the same experimental conditions (Figure 6).

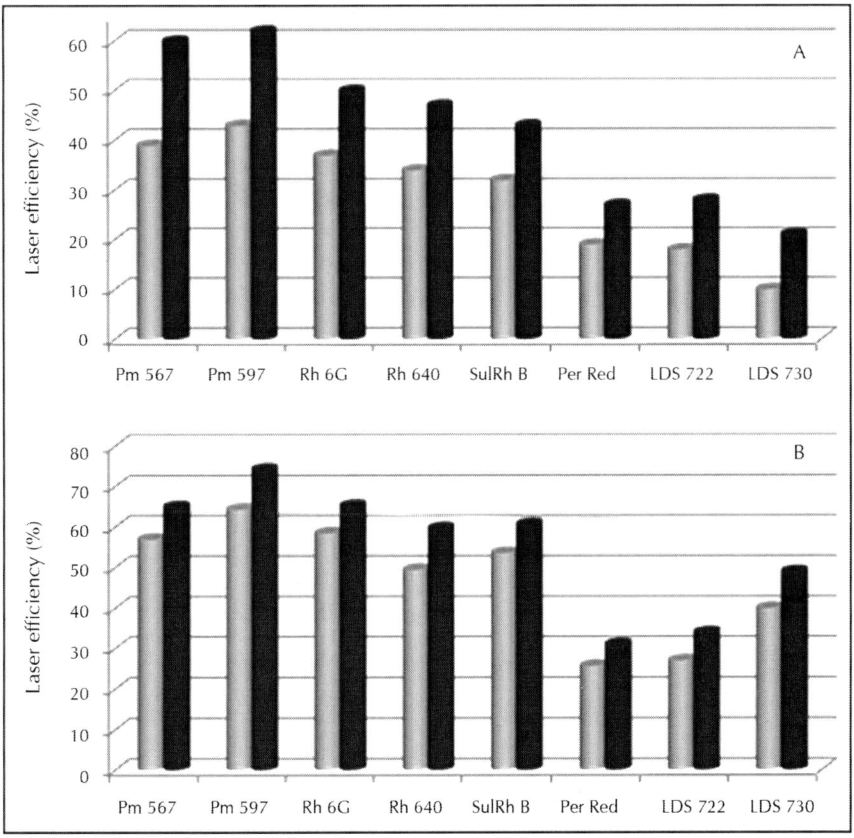

Figure 6. Laser efficiency of different dyes in the presence (black) and absence (grey) of 8MMAPOSS nanoparticles in liquid solution (a) and in solid samples (b), under transversal pumping at 532 nm with 6ns FWHM, 5.5 mJ pulses.

The solvents in the liquid solutions were ethyl acetate (PM567, PM597, Per Red), ethyl acetate: ethanol 1:1 (Rh6G, Rh640), ethyl acetate: ethanol 3:7 (SulfRhB), and ethanol (LDS722, LDS730). The composition of the polymeric matrices was MMA (PM567, PM597, Per Red), MMA:HEMA (2-hydroxyethyl methacrylate) 1:1 (Rh6G, Rh640, SulfRhB), and HEMA (LDS722, LDS730). In the solutions and solid matrices with POSS, the proportion of 8MMAPOSS nanoparticles was 13 wt %, which was the weight proportion of 8MMAPOSS which produced the best results. Dye concentration was 1.5×10^{-3} M (PM567), 4×10^{-4} M (Rh6G, SulfRhB, LDS722), 5×10^{-4} M (Rh640, PerRed), 6×10^{-4} M (PM597), and 8×10^{-4} M (LDS730).

The presence of 8MMAPOSS nanoparticles also improved remarkably the photostability of the laser emission, as illustrated in Figure 7, where is presented the actual evolution of the laser output of dye PM567 in a matrix of COP(MMA-8MMAPOSS 87:13) when the sample is pumped at 30 Hz repetition rate in the same position as compared with the evolution of the laser emission from the same dye incorporated into a matrix of pure PMMA pumped at lower repetition rate (10 Hz).

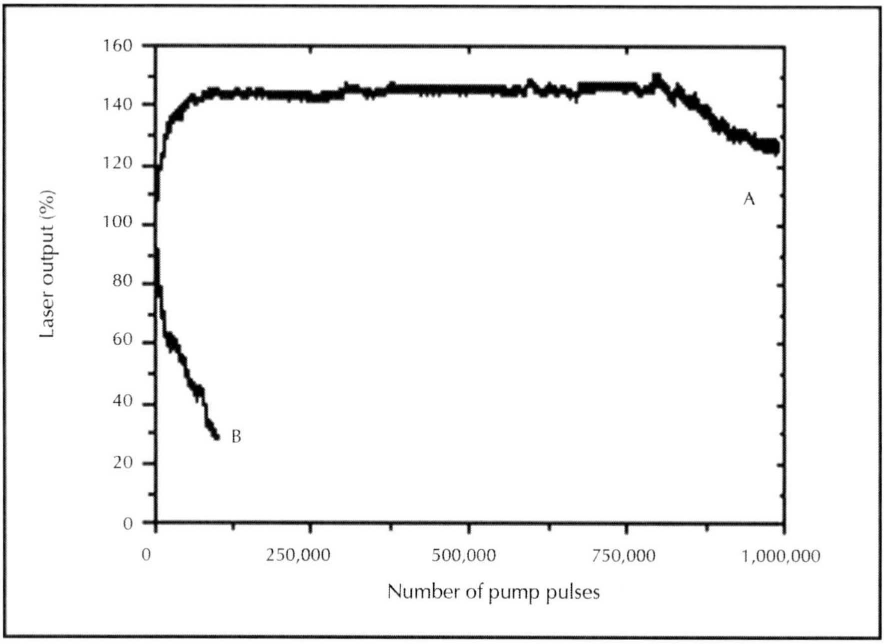

Figure 7. Normalized output as a function of the number of pump pulses in the same position of the sample for dye PM567 in: (a) copolymer (MMA-8MMAPOSS 87:13) pumped at 30 Hz, and (b) pure PMMA pumped at 10 Hz.

Trying to understand the process involved in the enhancement of the laser action induced by POSS, some photophysical and thermal properties of matrices containing PM567 were analyzed. It was found that the photophysics (molar absorption coefficient, absorption and fluorescence wavelengths, fluorescence quantum yields, lifetime, etc) of PM567 is no strongly affected by the presence of increased weight proportions of 8MMAPOSS. Thus, a change in the photophysics can be ruled out as being responsible for the exhibited lasing properties. Regarding the thermal properties, it is true that in the POSS-modified copolymers the thermal conductivity was enhanced from 0.182 W m^{-1} K^{-1} in pure PMMA to 0.2333 W m^{-1} K^{-1} for the materials with 50 wt % of 8MMAPOSS. However, this is neither the only nor the main cause of the improved lasing properties of the POSS-based materials, since other PM567-doped silicon-containing organic matrices (Costela et al., 2007) and sol-gel hybrid materials (García et al., 2008) with similar thermal conductivities exhibited a worse laser action when pumped under identical conditions.

Scattering has always been considered detrimental to laser action because it alters the direction and spatial coherence of the photons from the lasing mode of a conventional cavity. However, it has been demonstrated that, under suitable circumstances, the density of scattering particles allows an increase in the path length travelled by diffusive photons, and a large gain can occur in photoexcited media, which gives rise to scattering laser-like emission (Takeda and Obara, 2009). Thus, the question arises whether the POSS nanoparticles could act as passive weak scatterers when they are present in the dye-gain media in spite of their high optical homogeneity.

To assess this question in the present case, we characterized in detail the laser-induced spectroscopy of PM567 dye both in liquid and solid solutions with and without POSS nanoparticles, pumped under the same experimental conditions. The pump pulses were now incident on the samples at a 30° angle and the laser excitation energy was gradually increased from 0.9 to 6 mJ/pulse. The emission from the front-face of the sample was collected with an optical fiber, sent to an spectrometer with 0.1 nm resolution and detected with a CCD camera.

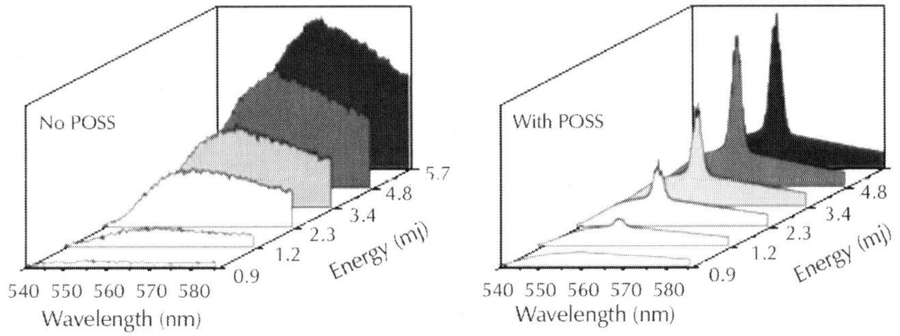

Figure 8. Front-face emission spectra as a function of the pumping energy at 532 nm from an ethyl acetate solution of PM567 dye that contains no POSS (left) or 13 wt % of 8MMAPOSS nanoparticles (right).

Figure 8 shows the evolution of the emission spectra from solutions of PM567 in ethyl acetate in the absence and presence of 8MMAPOSS nanoparticles. The excitation of a pure dye solution (no POSS), resulted in a broad photoluminescence spectrum, centered at 556 nm, which remained constant for the entire range of pumping energies. In the presence of 8MMAPOSS, even at the lowest (1%) weight proportion selected, a narrow stimulated emission band centered at 559 nm increased as the pump energy increased. Figure 8 (right) shows the evolution of the emission spectra from liquid solutions of PM567 in ethyl acetate with 13 wt % 8MMAPOSS at different pump energies above the threshold for the onset of amplified spontaneous emission (0.1 mJ). Similar results were obtained with the solid samples.

Increasing the density of scattering particles led to the broad tails of the photoluminescence being progressively suppressed and predominance of the gain-narrowed band. In the liquid phase, a multimode emission with narrow peaks with linewidth as small as 0.1 nm emerged on top of the globally narrowed spectrum. The wavelengths and heights of the narrow spikes in the emission spectra changed in a random fashion from one excitation pulse to another. The emission spectra were then completely uncorrelated from shot to shot since each excitation pulse illuminated a different configuration of scatters in a random walk motion inside the solution. On the contrary, in the emission spectra recorded from dye-doped solid-state POSS matrices, where the scattering particles were not moving and all other experimental conditions remained the same, narrow peaks were apparently lost and the sample front-face emission spectra became highly reproducible.

It could be argued that the parallel faces of the rectangular cell used in the laser-induced spectroscopy experiments described above could be contributing to provide feedback. To rule out this possibility, we performed some experiments using a triangular quartz cell, and obtained essentially the same results. What is more, when the triangular cell was pumped along the normal to the input face, unidirectional laser emission was detected (Figure 9) in the presence of 8MMAPOSS nanoparticles. No emission was detected in the absence of POSS nanoparticles in the solution.

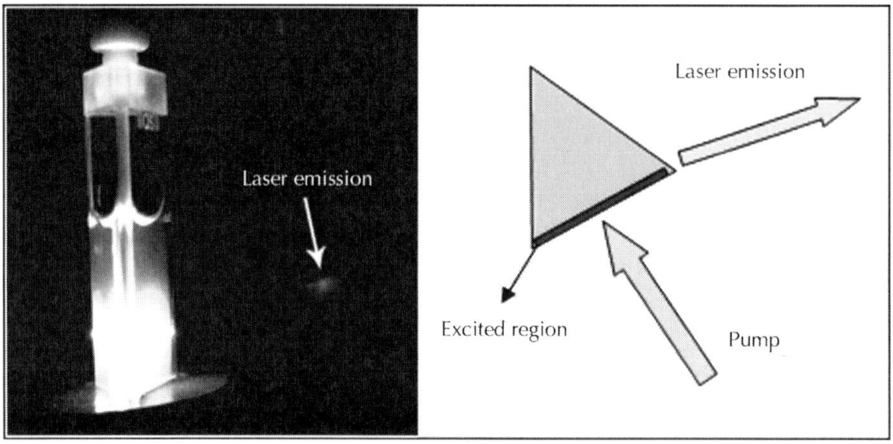

Figure 9. Emission from triangular cell filled with an ethyl acetate solution of PM567 with 13 wt % of 8MMAPOSS nanoparticles. Pumping at 532 nm with 5.5 mJ/pulse.

These results seem to point out to a NRFL mechanism where the dispersion of the POSS nanoparticles at a molecular level defines a highly homogeneous material that, when doped with laser dyes allow usual coherent laser emission but, in addition and in spite of their nanometer size, the POSS particles sustain an incoherent feedback into the coherent emission by multiple scattering. That is, the nanosized POSS particles allow a weak optical scattering of the emission that reinforces lasing by elongating the light path inside the gain media and provides an extra feedback. Because the scattering mean free path is much longer than the optical wavelength (Costela et al., 2009), the drastic spectral narrowing observed can be understood in terms of a light diffusion model with gain, where the phase of the light wave and interference effects are neglected (Wiersma and Lagendijk, 1996). According to this model, there is gain in the medium because of a mere increase in the path length of light in the region.

The above discussion allows concluding that that the remarkable improvement observed in the laser performance of dye-doped POSS systems is a direct consequence of scattering processes. Under defined experimental conditions, the dye-laser action is enhanced by stimulated emission build up in the gain direction by the elongated photon path length inside the gain media with scatterers. The laser-like emission adds up to the usual stimulated emission as long as the gain exceeds the losses in the scattering media. Once certain parameters (scatter density, sample size, pumping fluence) reach a critical value, the stimulated emission is effectively scrambled and the spectroscopically narrowed laser-like emission appears in all directions. Consequently, the optical losses that result from this leakage became larger than the optical gain, and the system becomes unstable with a decrease in intensity.

This is a new effect, where conventional directional lasing is enhanced by weak scattering in particles of nanometer size; that is, under some given conditions scattering is shown not to be detrimental to conventional laser emission but, on the contrary, it enhances that emission. It should be remarked that this is not random lasing, were the presence of scatterers in gain media results in omnidirectional laser emission with no spatial coherence but conventional directional laser emission enhanced via NRFL in weakly scattering particles.

Trying to understand the physics of the process, we developed a theoretical approach to model the behavior of gain media with very weak scattering in the absence of an external cavity, thus removing resonant feedback effects and distinguishing the true extent of the NRFL mechanism. We modeled the behavior of dye laser systems with nanometer-size scatterers (sub-diffusive regime: weak scattering centers in the Rayleigh limit) by using a rate equations formalism properly modified to consider weak scattering effects (Cerdán et al., 2011).

The geometry of the system will consist of an excited cylindrical region of length L and width $2r$ ($L >> r$). The emission grows in the direction defined by the axis of the cylinder, which is perpendicular to the pump direction. The emission properties will be evaluated at the exit of the excited region.

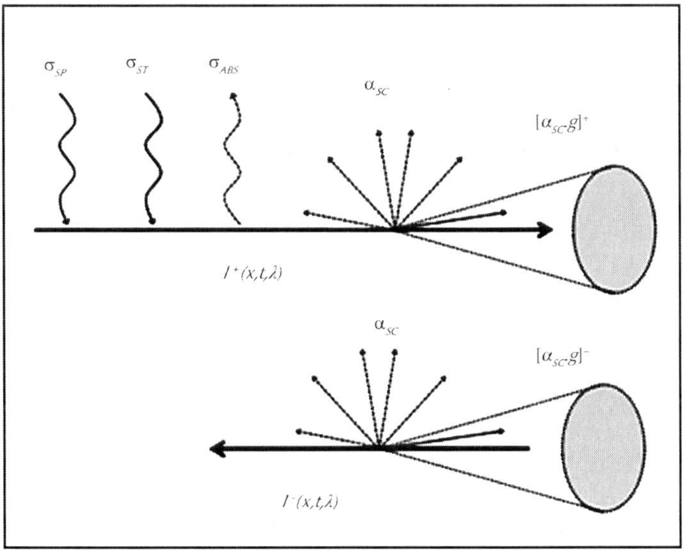

Figure 10. Processes considered in simulation: $I^+(x,t,\lambda)$ and $I^-(x,t,\lambda)$ forward and backward photon fluxes, σ_{SP} spontaneous emission, σ_{ST} stimulated emission, σ_{AB} losses due to absorption, σ_{SC} scattering losses, $[\alpha_{SC}.g]^+$ recovered forward scattering, $[\alpha_{SC}.g]^-$ recovered backscattering.

Figure 10 shows the scheme with the different processes taken into account in the simulation. Two counter propagating waves along the axis of the cylinder are considered, $I^{\pm}(x, t, \lambda)$. The contributions to the forward photon flux $I^{+}(x, t, \lambda)$ are: the spontaneous emission as seed, which is amplified by stimulated emission, reabsorbed by the ground state population, and dispersed due to scattering. The forward scattering of $I^{+}(x, t, \lambda)$ and the backscattering of $I^{-}(x, t, \lambda)$ within a certain solid angle are then restored to $I^{+}(x, t, \lambda)$. Since these scattering contributions only provide energy feedback, the system acts as a NRF laser. In the absence of scattering the system will emit amplified spontaneous emission (ASE).

Since, the size of the scatterers is much smaller than the scattered wavelength, the Rayleigh scattering formalism should be used. In this approach, the losses due to scattering are $\alpha_{SC}(\lambda) = \rho.\sigma_{SC}(\lambda)$, where ρ is the density of scatterers and $\sigma_{SC}(\lambda)$ is the wavelength dependent scattering cross section (Van de Hulst, 1981). For the sake of simplicity we consider the scatterer to be a dielectric sphere of radius r_S. Thus, $\sigma_{SC}(\lambda)$ can be calculated according to the equation (Wu et al., 2004):

The modified set of equations for the populations and photon fluxes reads:

$$\sigma_{SC}(\lambda) = \frac{8}{3}\left(\frac{2\pi r_S}{\lambda}\right)^4 \left(\frac{n^2-1}{n^2+2}\right)^2 \pi r_S^{\,2} \tag{1}$$

$$\begin{aligned}\frac{\partial N_1(x,t)}{\partial t} &= W(t)N_0(x,t) - \tau^{-1}N_1(x,t) \\ &\quad - N_1(x,t)\int \sigma_e(\lambda)[I^{+}(x,t,\lambda) + I^{-}(x,t,\lambda)]d\lambda \\ &\quad + N_0(x,t)\int \sigma_{01}(\lambda)[I^{+}(x,t,\lambda) + I^{-}(x,t,\lambda)]d\lambda\end{aligned} \tag{2}$$

$$N_0 + N_1 = N \tag{3}$$

$$\begin{aligned}\pm\frac{d}{dx}I^{\pm}(x,t,\lambda) &= \tau^{-1}N_1(x,t)E(\lambda)g^{\pm}(x) \\ &\quad + N_1(x,t)\sigma_e(\lambda)I^{\pm}(x,t,\lambda) \\ &\quad - \sigma_{01}(\lambda)N_0(x,t)I^{\pm}(x,t,\lambda) \\ &\quad - \alpha_{SC}(\lambda)I^{\pm}(x,t,\lambda) \\ &\quad + \alpha_{SC}(\lambda)I^{\pm}(x,t,\lambda)g^{\pm}(x) \\ &\quad + \alpha_{SC}(\lambda)I^{\mp}(x,t,\lambda)g^{\pm}(x)\end{aligned} \tag{4}$$

where $N_i(x, t)$ $(i = 0,1)$ is the population density (molecules' cm^{-3}) in the ground state S_0 and first excited (electronic) singlet manifold S_1, respectively, $W(t)$ is the pumping rate, $\sigma_{01}(\lambda)$ is the absorption cross section (cm^2) from S_0 to S_1 at wavelength λ, τ is the lifetime of the S_1 state in the absence of stimulated emission, $\sigma_{ST}(\lambda)$ is the stimulated emission cross section, given by $\sigma_{ST}(\lambda) = [\lambda^4 E(\lambda)]/8\pi c n^2 \tau$, where $E(\lambda)$ is the $S_1 \rightarrow S_0$ fluorescence spectrum normalized so that $\grave{o}E(\lambda)d\lambda = \phi$, ϕ being the quantum yield, and n is the refractive index of the dye solution. Finally, $g^{\pm}(x)$ are geometrical factors which determine a certain solid angle around the axial coordinate x (Figure 10), and are expressed as:

$$g^{+}(x)=\frac{1}{2}\left(1-\frac{L-x}{\left((L-x)^{2}+r^{2}\right)^{1/2}}\right) \tag{5a}$$

$$g^{-}(x)=\frac{1}{2}\left(1-\frac{x}{\left(x^{2}+r^{2}\right)^{1/2}}\right) \tag{5b}$$

where x is the position within the excited medium and. The plus and minus indices correspond to the solid angles subtended by the front and rear windows, respectively.

Triplet state population terms have been neglected due to the short pulse pumping conditions. In the excited state population equation (Equation (2)) the terms are described as follows. The first term describes the absorption of the pumping flux. The second term accounts for spontaneous decay via all available channels. The third term describes the depletion of the excited state due to stimulated emission of radiation in all wavelengths, and the last term describes the recovery of population due to self-absorption of the photon flux. The build-up of photon flux in the forward ($I^{+}(x, t, \lambda)$) and backward ($I^{-}(x, t, \lambda)$) directions is given by Equation (4), where $(d/dx)I^{\pm} = (\partial/\partial x)I^{\pm} \pm (\eta/c)(\partial/\partial x)I^{\pm}$. The right hand terms in the above Equation (4) are interpreted as follows. The spontaneous emission within the solid angle $g^{\pm}(x)$ around the axial coordinate x (first term) is amplified by stimulated emission (second term), and reabsorbed by the ground state population (third term). The fourth term takes into account the overall losses associated with scattering, while the fifth and sixth terms are the contributions of the forward scattering of $I^{+}(I^{-})$ and the back scattering of I^{-} (I^{+}) within the solid angle $g^{+}(g^{-})$, respectively. These contributions are expressed as the total amount of flux in the corresponding direction lost due to scattering weighted by the corresponding solid angle. In order to facilitate the calculations we have assumed that the scattering is isotropic, that, for unpolarized light and for the solid angles involved, is a reasonable approximation (Van de Hulst, 1981).

$$g^{+}(x)=\frac{1}{2}\left(1-\frac{L-x}{\left((L-x)^{2}+r^{2}\right)^{1/2}}\right)$$

Table 1. Parameters used in the computation.

Parameter	Numerical Value
τ (ns)	5.78 [a]
ϕ	0.63 [a]
η (ethyl acetate)	1.37

Table 1. *(Continued)*

Parameter	Numerical Value
L (cm)	1.4
r (μm)	150
λ_p (nm) (pump wavelength)	532
N (cm^{-3})	9′10^{-17} (1.5 mM)
Scattering losses (cm^{-1}) [α_{SC}]	0-0.8
Pump energy (mJ) [E_{pump}]	0.1-25
Pump pulse FWHM (ns) [τ_p]	5 (Gaussian pulse)
[a] Ref. 40	

The parameters needed for the computation are given in Table 1. They correspond to a solution of dye Pyrromethene 567 (PM567) in ethyl acetate incorporating generic. $E(\lambda)$ and $\sigma_{01}(\lambda)$ were obtained from curves similar to those in Figure 2 of ref. (Costela et al., 2002) (In Ref. [40], we studied the photophysics of PM567 and other dyes in six different solvents and only published as an example the curves corresponding to solutions of PM567 in methanol. For the calculations here reported the unpublished but similar curves of PM567 in ethyl acetate have been utilized). The excited region is pumped transversely with 5ns Gaussian pulses at 532 nm with different pump energies. As boundary conditions we have used $I^+(x = 0, t, \lambda) = I^-(x = L, t, \lambda) = 0$, with initial conditions $I^+(x, t = 0,\lambda) = I^-(x, t = 0, \lambda) = N_1(x, t = 0) = 0$ and $N_0(x, t = 0) = N$, where N is the density of dye molecules.

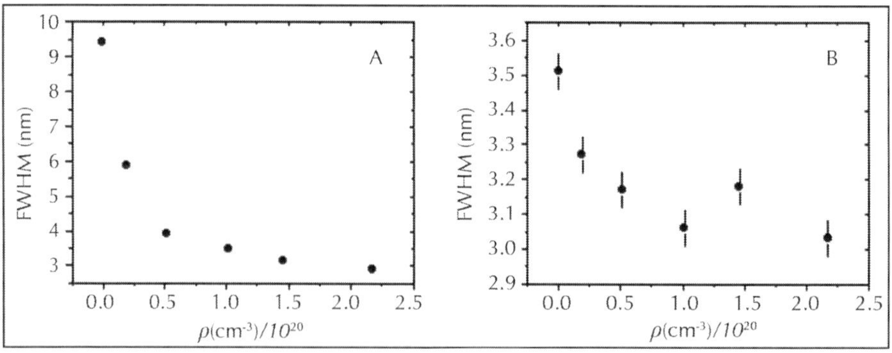

Figure 11. Calculated (a) and experimental (b) FWHM of the emission for different densities of nanoparticles.

Spectral narrowing of the emission and an increase of the laser output energy due to the presence of the nanoscatterers are predicted by the model in good qualitative agreement with experiment. In Figure 11 it is shown the calculated (a) and experimental (b) evolution of the emission bandwidth for different densities of scatterers. It is clearly seen an additional collapse in the linewidth of the emission caused by the presence of the nanoscatterers as compared with the case without losses.

In Figure 12 is presented the calculated (a) and experimental (b) relative output energy improvement, defined as $\Delta E = (E_{OUT} - E_{OUT}^{0})/E_{OUT}^{0}$ ′ 100%, where E_{OUT} and E_{OUT}^{0} are the output energy with and without losses, respectively. The inset in Figure 12 (a) covers the region of physically available 8MMAPOSS densities, to be compared with the experimental results in (b). The efficiency improvement ΔE increases as the scattering losses are increased, and this improvement is only possible is scattering feedback is taken into account.

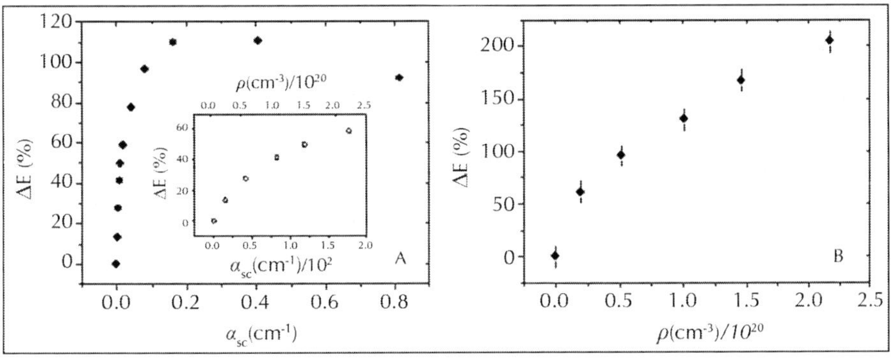

Figure 12. Calculated (a) and experimental (b) relative output energy improvement of the emission for different densities of nanoparticles.

Thus, the modified rate equations formalism shows that the presence of weak scattering in the gain medium, instead of being detrimental to laser efficiency, can lead to an increase in the output energy and to narrower emission linewidths, without much affecting the beam divergence.

The proposed model provides a theoretical explanation of the observed enhanced conventional laser emission in gain media with scattering centers and establishes the limits for the region where gain dominates over scattering losses. The mechanism is independent from the nature of the scattering centers or the medium, working both in liquid and solid solutions. A critical factor in the physics of the emission is the nanometric size of the scattering particles.

Laser Emission from Mirrorless Waveguides

We have also investigated the effect of the presence of POSS in photosensitized planar waveguides. Thin films with thickness between 2 and 9 μm of dye-doped organic gain media incorporating 8MMAPOSS nanoparticles were deposited onto glass substrates, pumped transversely, and their edge-on emission collected (Cerdán et al., 2009).

The presence of any amount of 8MMAPOSS in the thin film composition resulted in a spectral narrowing and red shift of the ASE emission from the samples (Figure 13(a)). In films with 5.5 μm thickness, when the content of 8MMAPOSS in the polymer increases to 50 wt % proportion, multimode emission with narrow peaks in top of a globally narrowed ASE spectrum does appear (Figure 13(b)), the intensity of the emission increases by more than tenfold, and in the center of the previous fringe shaped emission does appear a bright spot (Figure 13(c)). The linewidth of the peaks in Figure 13(b) is smaller than 0.1 nm, which is the resolution of our detection system.

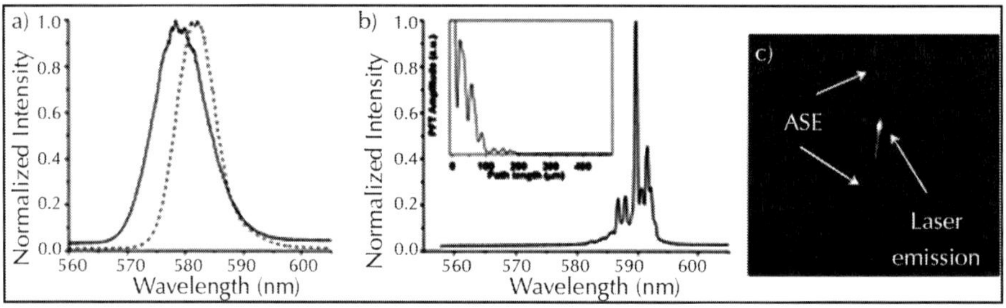

Figure 13. (a) Emission spectra from 5.5 μm thick films containing 0% (solid line) and 20% (dashed line) 8MMAPOSS; (b) Emission spectrum from a 5.5 μm film containing 50% 8MMAPOSS and (inset) Power Fourier Transform of spectrum;and (c) Edge-on emission from film in (b).

In Figure 14 we have represented the evolution of the intensity of the main narrow peak in the emission from the sample containing 50 wt% 8MMAPOSS (Figure 13(b)) with the pump intensity. A clear pump threshold for the narrowband emission to appear is evident in the Figure 14. Taken together, the narrow linewidth of the peaks in the multimode emission, the small divergence of the emission, and the existence of a pump threshold are evidences of the laser nature of the narrow emission lines.

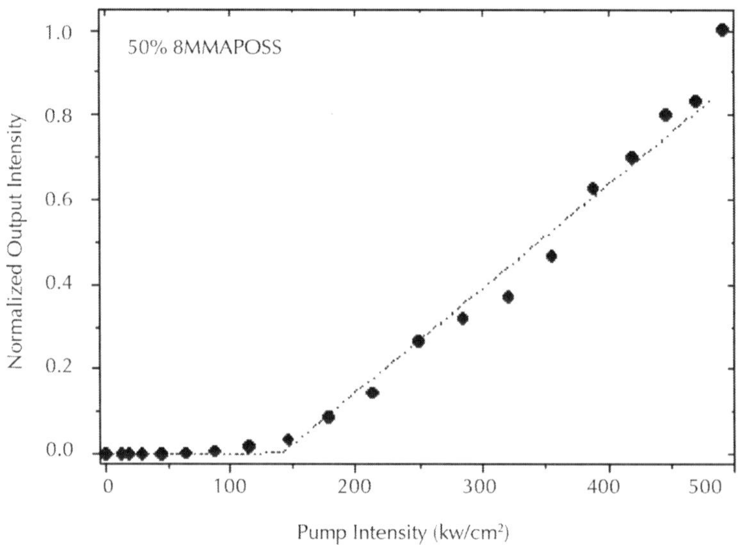

Figure 14. Dependence on pump intensity of the intensity of the main peak in the multimode emission from a 5.5 μm thick film with 50 wt% 8MMAPOSS.

Another feature of this multimode laser emission is the randomness of the peak wavelengths when moving the sample. This leads us to ascribe the multimode emission to a mechanism of coherent random lasing (Takeda and Obara, 2009). That is, the distribution of scatterers within the excited region, that conforms to random ring cavities, changes in a random fashion when the sample is moved, leading to different oscillation conditions (random cavities) in each sample region. It is possible to gain insight into the excited random cavities by calculating the Power Fourier Transform (PFT) of spectra such as that in Figure 13(b). The PFT of the emission spectrum (in $k = 2\pi/\lambda$ space) from a well-defined laser cavity shows peaks at Fourier components $p_m = mL_C n/\pi$, where m is the order of the Fourier harmonic, L_C is the cavity path length, and n is the refractive index of the gain medium (In Ref. [40], we studied the photophysics of PM567 and other dyes in six different solvents and only published as an example the curves corresponding to solutions of PM567 in methanol. For the calculations here reported the unpublished but similar curves of PM567 in ethyl acetate have been utilized). The inset in Figure 13(b) shows the calculated ensemble-averaged PFT spectrum of the emission spectrum shown in Figure 13(b). The first sharp peak in the PFT spectrum corresponds to the fundamental Fourier component $m = 1$, and gives a mean cavity path length $L_C \approx 45$ μm (assuming $n = 1.4977$).

Summarizing the above discussion, we ascribe the spectral narrowing and red shift, such as those appreciated in Figure 13(a), to an effect of NRFL present in weakly scattering systems, where diffusive photos experience multiple scattering. Each scatterer contributes incoherently to this feedback, elongating the light path inside the gain medium and, above a certain pump threshold, results in an additional spectral narrowing. According to this, the higher the scatterer density (8MMAPOSS content) is the narrower the emission spectrum results. The red shift in the emission with 8MMA-POSS content can be understood as being due to two related effects: on the one hand, the self-absorption of the emitted radiation is higher due to the increased photon path, and on the other hand, the scattering losses are higher for shorter wavelengths.

CONCLUSION

Under defined experimental conditions, dye laser action in media incorporating nanoparticles is enhanced by stimulated emission build up in the gain direction by the elongated photon path inside the medium. This laser-like emission adds up to the usual stimulated emission as long as the gain exceeds the losses in the scattering media.

A rate equation formalism is used to model the behavior of gain media with weakly scattering nanoparticles. Spectral narrowing of the emission and an increase of the laser output energy due to the presence of nanoscatterers is predicted by the model and observed experimentally, showing that scattering is not always detrimental to conventional laser emission as thought so far. This paradoxical behavior is explained in terms of NRFL, where the photon path enlargement caused by multiple scattering provides an extra feedback and enhances incoherently the magnitude of the amplification process. The model provides a theoretical explanation of the observed enhanced conventional laser emission in gain media with scattering centers and establishes the limits for the region where gain dominates over scattering losses. The mechanism is independent from the nature of the scattering centers or the medium, working both in liquid and solid solutions, as long as the size of the scatterers is nanometric.

When the gain medium incorporating nanoparticles is deposited onto glass substrates defining a planar asymmetric slab waveguide, multimode laser emission can appear without the need of the presence of any resonant substructure.

KEYWORDS

- **Methyl methacrylate**
- **Poly(methyl methacrylate)**
- **Polyhedral oligomeric silsesquioxanes**
- **Power Fourier transform**
- **Pyrromethene**
- **Solid-state dye lasers**

ACKNOWLEDGMENTS

This work was supported by Project MAT2007-65778-C02-01 of the Spanish CICYT. The V. M. and M. E. P.-O. thank Spanish CSIC for their JAE-Doc postdoctoral and JAE predoctoral contracts, respectively. The L.C. thanks MICINN for a predoctoral scholarship (FPI, cofinanced by Fondo Social Europeo).

Chapter 10

Synthesis and Characterization of Zinc Sulfide Nanocrystals and Zinc Sulfide/Polyvinyl Alcohol Nanocomposites for Luminescence Applications

Meera Ramrakhiani and Vikas Nogriya

INTRODUCTION

Zinc sulfide (ZnS) nanoparticles and zinc sulfide/polyvinyl alcohol (ZnS/PVA) composite films of different crystalline sizes have been synthesized by chemical route. The XRD study shows formation of ZnS nanocrystals, having size between 2–4 nm. ZnS nanocrystals exhibit cubic zinc blende crystal structure, while the ZnS/PVA nanocomposite films show wurtizite crystal structure with hexagonal phase. Optical absorption spectra show the blue shifted absorption edge. Photoluminescence (PL) was excited by three different wavelengths. The PL studies of ZnS nanoparticles show the three peaks at 450, 480, and 525 nm and the PL spectra of ZnS/ PVA nanocomposite film show highly broadened emission band in violet–blue region. The emission may be attributed to the presence of sulfur vacancies in the lattice. The PL intensity is found to increase by reducing the particle size. The electroluminescence studies of ZnS nanoparticles and ZnS/PVA nanocomposite films show lower turn on voltage and higher brightness for smaller nanoparticles. The electroluminescence spectra of ZnS/PVA nanocomposite films show violet-blue light emission with two peaks at 425 and 480 nm.

Semiconductor nanoparticles or quantum dots have received intensive attention in recent years since they exhibit size dependent properties useful for various electronic and optoelectronic applications (Alivisatos et al., 1989; Brus, 1984; Efros and Efros, 1982; Everett, 1988; Jaehun et al., 2002; Kale et al., 2006; Kubo et al., 2002; Matijevic, 1986). Since these nanoparticles have very high surface to volume ratio, surface defects play an important role in their properties. The ZnS is a direct wide band gap semiconductor that is one of the most important materials in photonics. This is because of its high transmittance in the visible range and its high index of refraction (about 2.2). The ZnS doped with manganese (Mn) exhibits attractive light-emitting properties with increased optically active sites for applications as efficient phosphors. The ZnS has a band gap of 3.68 eV at 300K with exciton Bohr diameter 5.2 nm. This corresponds to ultra violet radiation for optical interband transition with a wavelength of 340 nm. Wide band gap semiconductors are ideal materials for studies of discrete states in the gap. Visible luminescence can only originate from transition involving these localized states. The ZnS in common with most (but not all) inorganic phosphors, must be "activated" by suitable impurities in order to produce luminescence.

It is worth noting some of the applications that motivated the studies of ZnS. The ZnS has many applications. Rutherford and others first used ZnS in the early years of nuclear physics as a scintillation detector. A scintillation detector measures ionizing radiation. The ZnS does this because it emits lights on excitation by x-rays or electron beam. Utilizing this aspect, ZnS is used in x-ray screens and cathode ray tubes. The ZnS is also an important phosphor host lattice material used in electroluminescent devices (ELD), because of the band gap large enough to emit visible light without absorption and the efficient transport of high energy electrons. The ZnS exists in two forms: a low-temperature cubic (sphalerite or zincblende) structure and a high-temperature hexagonal (wurtzite) structure. The sphalerite structure can be derived from a cubic close packing of ions, while the wurtzite structure is derived from a hexagonal close packing scheme.

Figure 14.1 shows each crystal structure. It has been reported that electroluminescent material must contain both sphalerite and wurtzite phases (Arterton et al.,1992). The wurtzite type structure predominates when the bonding is primarily ionic whereas the more covalent systems favor the sphalerite form. The cubic phase of ZnS is not grown as easily as the hexagonal phase, thus making the hexagonal phase more appealing for EL device applications (Bellotti et al., 1988).

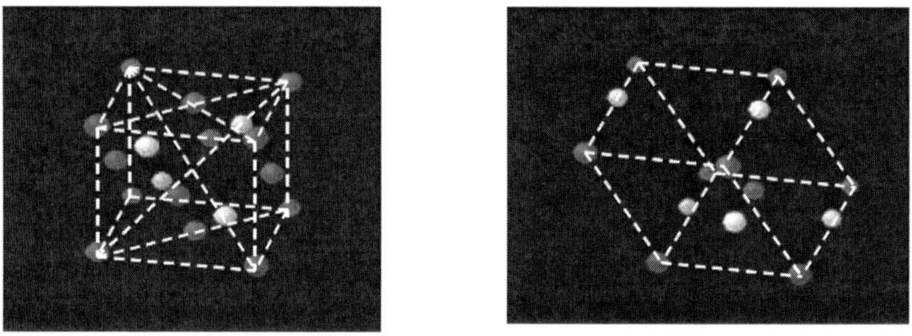

a) b)

Figure 1. ZnS crystal structure (a) zinc blende and (b) wurtzite.

During the last years, increasing attention has been devoted to nanocomposites, for example a transparent polymer matrix with inorganic nanoparticles embedded for their many interesting applications in optoelectronics (Colvin et al., 1994; Huynh et al., 2002; Tessler et al., 2002). To fabricate such a nanocomposite, the nucleation of nanoparticles directly in the polymer matrix can be useful if it is possible to control the nanoparticle dispersion and size within the polymer matrix, thus suppressing segregation. As a new class of optoelectronic materials, nanocomposites consisting of semiconductor nanocrystals and organic polymers are attracting the interest of many workers (Wang and Herron 1992).

In the present chapter, we report the synthesis of ZnS nanocrystals and ZnS/PVA nanocomposite films, their structural and optical characterization and photo- as well as electro-luminescence investigations.

SYNTHESIS

There have been many efforts to synthesize size selected ZnS with a very narrow size distribution, however only a few were successful. Most of the techniques follow a capping route and a wet chemical synthesis with solvents like mercaptoethanol, ethanol, methanol, acetonitril, dimethylformamide (DMF) etc. Verities of physical and chemical techniques are used for the synthesis of nanoparticle polymer composite, so that the film can be cast directly, thus avoiding the loading of the particles via multistage processes. In present studies the nanocrystaline ZnS samples in powder form and in nanocomposite form were prepared by chemical technique.

Preparation of Zns Nanoparticles

It is essential to restrict the size distribution in nanocrystalline samples. In order to reduce the agglomeration of the nanocrystallites, the crystallites are generally capped with organic materials (Manzoor et al., 2004; Nanda et al., 1999; Rosseti et al., 1985) such as thiophenol, mercaptoethenol, ethylene glycol, mercapto-acetic acid, etc. The two salt solutions, each having one component of the binary semiconductor to be synthesized, are mixed in the presence of capping agent. The concentration of the solutions, especially that of capping agent, controls the particle size. The ZnS nanoparticles have been prepared by mixing 0.01 M solutions of zinc chloride ($ZnCl_2$) and sodium sulfide (Na_2S) in presence of different concentrations of mercaptoethenol (C_2 S_5 OSH) capping agent varying from 0 to 0.02 M.

All the chemicals were of analytical grade and used as received without further purification. All the experiments were conducted under ambient atmosphere. The aqueous solution of mercaptoethenol was first added drop wise in the 0.01M solution of ZnC12 with the help of a burette, at the rate of 1 ml/min while stirring the solution continuously. Magnetic stirrer was used for stirring the solution. Thereafter, 0.01M solution of Na2S was mixed drop wise into the solution. Subsequently, a milky color solution was obtained. This solution was kept for 24 hr. White precipitate settled down in the bottom of the flask. This precipitate was removed and washed several times with the double distilled water. In this reaction, sodium chloride (NaCl) was formed which was removed by washing the solution. The unreacted mercaptoethenol and Na2S are removed also. The reaction can be summarized in the equation form:

$$ZnCl_2 + Na_2S \xrightarrow{C_2H_5OSH} ZnS + 2NaCl$$

This washed solution was centrifuged. Finally, the precipitate was spread over a glass substrate and air–dried at room temperature. The capping agent C_2H_5OSH restricts the particle size. For the present investigation, a number of samples were prepared by using different concentrations of capping agent varying from 0 M to 0.02 M as given in Table 14.1.

Table 14.1. The concentration of various compounds used for preparing the ZnS nanoparticles.

Sample Name	C_2S_5OSH concentration	Volume of C_2S_5OSH (in 100 ml)	Weight of $ZnCl_2$ in 100 ml (0.01M)	Weight of Na_2S in 100 ml (0.01M)
ZnS I	0.000M	Nil	136.2 mg	78.02 mg
ZnS II	0.005M	0.0345	136.2 mg	78.02 mg
ZnS III	0.010M	0.069	136.2 mg	78.02 mg
ZnS IV	0.015M	0.103	136.2 mg	78.02 mg
ZnS V	0.020M	0.138	136.2 mg	78.02 mg

Preparation of ZnS/PVA Nanocomposite Film

For the synthesis of composite films, 400 mg PVA was dissolved in DMF by constant stirring and heating at 70°C. After proper solution was obtained, zinc acetate was added to it in appropriate quantity, so that ZnS loading in polymer is 2, 5, 10, 20, 30, and 40% by weight (given in Table 14.2).

Table 14.2. The weight of various compounds used for preparing ZnS/PVA composite films.

S. No.	Sample Name	Weight of PVA	Weight of Zinc Accetate	Loading of ZnS
1	A	400 mg	8 mg	2% ZnS
2	B	400 mg	20 mg	5% ZnS
3	C	400 mg	40 mg	10 % ZnS
4	D	400 mg	80 mg	20% ZnS
5	E	400 mg	120 mg	30% ZnS
6	F	400 mg	160 mg	40% ZnS

The resulting solution was stirred for 30 min and concentrated to 10 ml. The solution was refluxed by applying nitrogen and then hydrogen sulfide (H_2S) gas was passed for a 30 second to this clear solution of zinc acetate and PVA. The solution immediately turned milky white. Now again the solution was stirred for a few seconds, then caste over glass slides and conducting glass plate also and dried in oven in a mercury pool to obtain uniform film of ZnS/PVA nanocomposite. The polymer film acts as a stabilizer-cum-matrix, resulting in a high uniformly in the size of nanocrystals. In this work commercially available PVA was used, DMF is used as solvent, it is good dispersion medium. Here, in the sample, the content of Zn is increased with higher loading of ZnS, and the H_2S gas was passed for fixed 30 seconds, which is sufficient to react with available Zn ions.

The chemical mechanism of the preparation of ZnS nanoparticles in PVA matrix is shown in Figure 2. From the preparation method used here, there is a high probability of free Zn^{2+} ions at the polymeric chain without S^{2-} ion counterparts. Zinc accetate dissociate into zinc ions (Zn^{2+}) and acetate (Ac^-) ions in aqeous solution. Similarly, H_2S dissociate into its respective cations and anions (Manzoor et al., 2003; Sharma and Bhatti, 2007). Particles of ZnS nucleate due to the reaction between Zn^{2+} and S^{2-}, which subsequently grow by consuming more ions from the solution. Upon nucleation, the surface energy of the particle is very high and consequently the surface is passivated by cations towards the surface of the particles. The Zn^{2+} reacts with S^{2-} and get incorporated into the crystal lattice of the nucleus. To avoid agglomeration, a repulsive force must be added between particles to balance the attractive force. This is achieved by absorbing a layer of polymer over nanoparticles inducing steric hinderence by empolying PVA as the stabilizing agent. High viscosity of polymer restrict the particle growth.

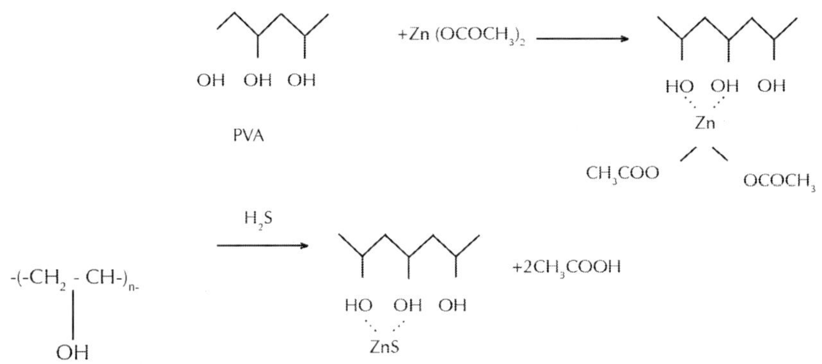

Figure 2. (a) Molecular Structure of PVA and (b) Chemical mechanism of ZnS/PVA composite film.

CHARACTERIZATION

In order to understand the inter-relationship between structure and properties crystalline materials need to be characterized on both atomic and nanometer scales. The characterization of above involves determining the shape and size of nanoparticle and understanding the inter-particle interaction. This information is important both from the scientific and the industrial application point of view. In present work, transmission electron microscopy (TEM) and X ray diffrection (XRD) technigues have been used to characterize the ZnS samples.

Transmission Electron Microscopy

In order to investigate the particle size, the samples were analyzed by TEM. Figure 14.3 (a) shows a TEM image of the ZnS nanocrystal synthesized with 0.01 M capping agent concentration. It shows the agglomerate of a few nanometer length. The particles are found to be speherical with an average size 3 nm for ZnS III sample and this image indicate formation of nanoparticles with multimodal distribution of shape and size. The selected area electron diffrection (SAED) pattern of sample consist of broad diffuse rings, which are indicative of the small size of the particles. (Figure 14.3(b)), These electron diffrection (ED) pattern clearly reveal the develpoment of lattice ordering formed in the nanocrystals. The ED pattern indicate that the closed packed plane is perpendicular to the field of veiw. The diffraction rings can be indexed to the (111), (220), and (311) plane confirming the cubic structure.

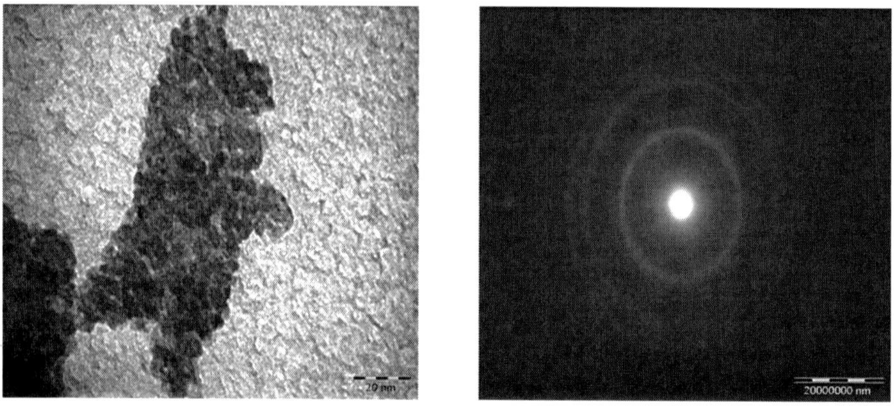

Figure 3. (a) TEM image of ZnS nanocrystalline sample and (b)SAED pattern of ZnS III sample.

XRD of ZnS Nanocrystals

The X ray diffraction studies have been undertaken for all the samples in order to determine their crystal sturcture, and crystallite size. The XRD pattern have been obtained by Rigaku Rotating X Ray Diffractrometer with irradiation from copper Kα line (λ= 1.5418 Å). Figure 4 shows the diffrection patterns of ZnS nanocrystals prepared with different mercepteoethanol capping agent concentration. Three major peaks at 2θ =28.8, 48.3, and 57.2° are obtained in all the samples. The phase identification was carried out with the help of standered JCPDS data (file No 80-0020), which indicate that all the samples have cubic phase with zinc blende sturcture. Diffrection peaks from (111), (220), and (311) phanes have only appered in XRD patterns and all other high angle peaks have submerged in the background due to the nanosize of the particles. The lattice spacing 'd' is calculated from the Bragg's formula

$$d = n\lambda/2 \sin\theta \tag{1}$$

The average crystal size (D) has been calculated using Scherrer's formula (Guinier, 1963) based on the full width at halh maximum (FWHM) of the prominent peaks due to (111) plane.

$$D = \frac{K\lambda}{\beta \cos\theta} \tag{2}$$

where D is the crystal size, β is full width at half maxima (FWHM) of X-ray refelection expresed in radians and θ is the position of the diffraction peak in diffractrograms. The mean crystal size of ZnS nanocrystal is given in Table 3. It is seen that as expected, the diffraction patterns of nanoparticles are broandened and particle size is reduced as the capping agent concentration is increased.

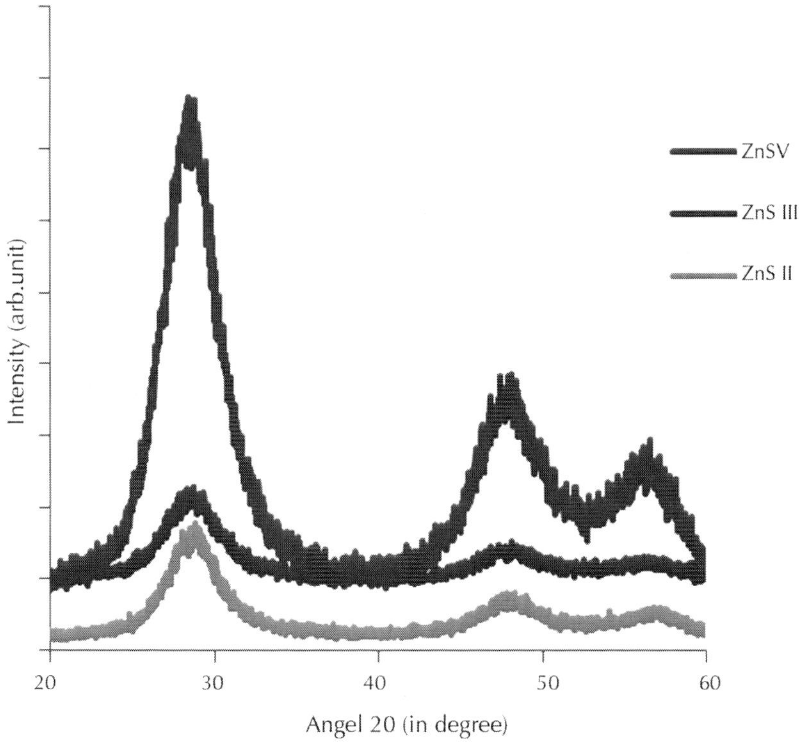

Figure 4. XRD pattern of ZnS nanocrystals.

Table 14.3. Analysis of X-Ray diffraction pattern of ZnS nanocrystals.

Sample name	Capping agent concentration	Angle 2θ (in degree)	Hkl	d (in Å)	Standarad 'd'(in Å) (JCPDS-80-0020)	Lattice constant a (in Å)	Crystal size D (in nm)
ZnS II	0.005 M	28.8	111	3.08	3.08	5.33	3.8 nm
		48.3	220	1.88	1.88	5.31	
		57.2	311	1.60	1.61	5.30	
ZnS III	0.01 M	28.72	111	3.10	3.08	5.38	3.2 nm
		48.3	220	1.88	1.88	5.31	
		56.75	311	1.62	1.61	5.38	
ZnS V	0.02 M	28.52	111	3.12	3.08	5.40	2.5 nm
		48.12	220	1.89	1.88	5.34	
		56.56	311	1.62	1.61	5.37	

The lattice constant 'a' of the ZnS nanocrysta is calculated using Eq.(3). The calculated lattice constant 'a' is given in Table 3, which is well mateched with standarad lattice constant 'a' (a= 5.34). The interplanner distance 'd' is also matched with standard 'd' of JCPDS data card.(JCPDS-80-0020).

$$a = \frac{\lambda}{2\sin\theta}\sqrt{\left(h^2 + k^2 + l^2\right)} \quad (3)$$

XRD of ZnS/PVA Composite Film

The shape, size and crystal structure of ZnS/PVA nanocomposite films have been studied by XRD at IIT Delhi using X-Pert thin film X ray diffractomrter with Cu target operated at 45 KV and 40 mA. The data collected in range of 10 to 60° in the intervals with step size 0.03. Figure 5 represent the three diffreent spectra of ZnS/PVA composite with different loading of ZnS in ZnS/PVA matrix. XRD pattern shows a peak at 2θ = 19.5° and 40.2° due to PVA (Hong et al.,1998), while a shallow shoulder was formed corresponding to the amorphous part of the polymer behind the diffrction peak angle at 2θ = 19.5° at all the samples. Small peaks are observed at 28.2°, 39.5°, 47.4°, and 52.3° due to ZnS. The phase identification was carried out by JCPDS data File No(JCPDS-36-1450). It can be seen that the XRD pattern of ZnS/PVA composite thin film can be consistently indexed on the basis of wurtzite structure, in which the lattice planes corresponds to (002), (102), (110) reflection of hexagonal phase.The obtained d values were compared with the powder diffrection data file no(JCPDS-36-1450) .

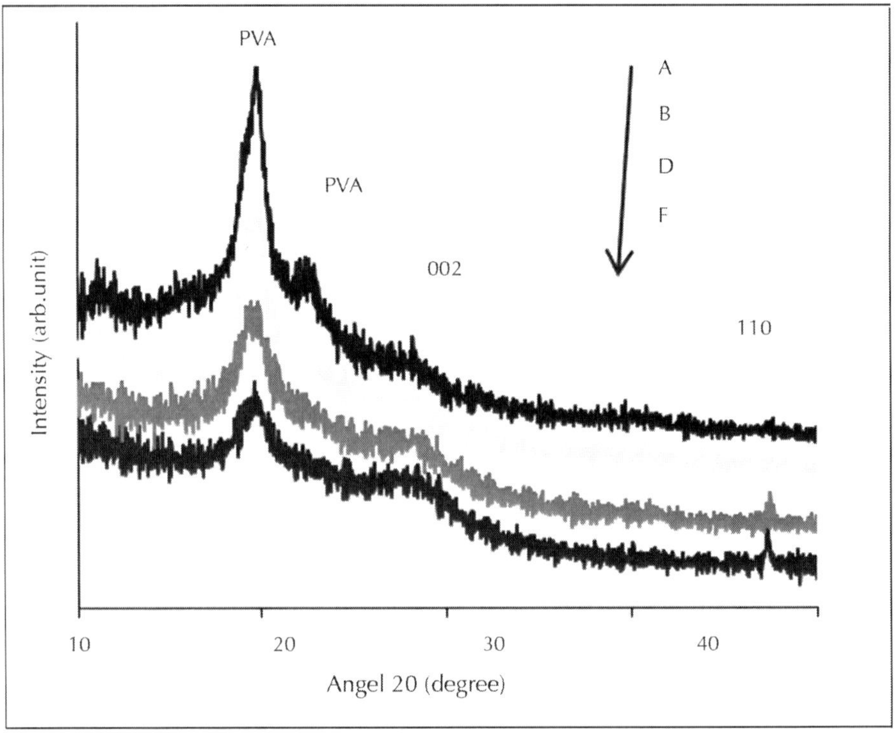

Figure 5. XRD pattern of ZnS/PVA nanocomposite film.

Table 4 shows the d- values obtained from XRD using Eq.(1). The lattice parameters 'a' and 'c'were determined for hexagonal structure by using the following relations (Roth, 1967):

$$\sin^2\theta = \left[\frac{\lambda^2}{3}\left(\frac{h^2+hk+k^2}{a^2}\right)\right] + \frac{\lambda^2 l^2}{4c^2} \tag{4}$$

$$C_{hex} = (1.633)a_{hex} \tag{5}$$

The calculated value of lattice constant 'a' and 'c' are given in Table 14.4, which are nearly same as literature, a=3.82 and , c= 6.25 for hexagonal ZnS wurtzite sturcture. Rincon et al. have also reported the hexagonal wurtzite structure with some low intensity peaks of ZnO (zincite) for ZnS films (Kayanuma, 1988).

Table 14.4. Analysis of X-Ray diffraction pattern of ZnS/PVA nanocomposite film.

Sample name	Loading of ZnS	2θ (degree)	hkl	Inter planner spacing 'd'	Standard'd' (JCPDS-36-1450)	Lattice constant 'a' (Å)	Lattice constant 'c' (Å)	Crystal size D (nm)
B	5 %	28.3	002	3.14	3.12	3.85	6.29	5.8 nm
		39.7	102	2.26	2.27	3.82	6.24	
D	20 %	28.5	002	3.12	3.12	3.82	6.24	4.7 nm
		47.7	110	1.90	1.91	3.81	6.22	
F	40 %	28.9	002	3.08	3.12	3.78	6.16	3.2 nm
		39.9	102	2.25	2.27	3.80	6.21	

From the figure, it is observed that when the loading of ZnS increases, FWHM increses and peak is shifted to higher diffrection angle with decreasing crystal size. The grain size of the crystallite was determined from FWHM of the intense peak corresponding to (002) plane by using Scherrer formula equation. The values of crystal size with different loading of ZnS are given in Table 14.4.

From the table it is clear that the crystal size decreases with higher loading of ZnS in PVA matrix. A slight increase in diffrection angle is may be a result of lattice contraction expected to occure because of high surface to volume ratio (Rincon et al., 2003) but further confirmation is required.

OPTICAL ABSORPTION

The absorption spectra show a clear picture of energy levels, density of states and allowed transitions in the materials. In case of nanocrystallites, the electrons, holes and excitons have limited space to move and also limited energy states. Thus, their energy spectrum is quantized. As the size of crystal is decreased below Bohr exciton size, the electronic states are descretized and result in widening of band gap and increase the oscillator strength. The phenomena of radiation absorption in a material is considered to be due to inner shell e^-, valence band e^-, free carriers and electron bound to localized impurity centers. If the energy of incident photon is less than the energy band gap of any material then this photon energy will not be observed and material is transparent to those wavelengths. As the photon energy increases to the band gap or greater than band gap energy, then photons are absorbed and sudden increase in absorption is obtained at these wavelength. The ZnS is a large band gap material, which has the band gap energy 3.68 eV at 300K. Optical absorption spectra of the ZnS nanoparticles and nanocomposites samples were studied by Perkin Elmer λ –35 UV- Visible spectrometer in the range 200 nm to 800 nm. It scans with speed- 240 nm/min and data interval 1 nm.

Absorption Spectra of ZnS Nanoparticles

The ZnS nanocrystals were dispersed in Dimethyl Sulfono Oxide (DMSO) and spread over glass plates and then heated to 60°C in order to obtain a layer of the ZnS nanocrystals on the glass plates. These were used for obtaining absorption spectra. The glass plate was taken as reference. Figure 14.6 shows the UV/VIS optical absorption spectra for ZnS–II, ZnS–III, ZnS–IV, and ZnS–V samples prepared with capping agent concentration of 0.005M, 0.01M, 0.015M, and 0.02M, respectively. It can be seen from the spectra that there is practically uniform absorption in the visible range (800–390 nm). Absorption increases suddenly in the UV region. No absorption peaks are found. For ZnS II, sample sudden increase in the absorption occurred at 300 nm similarly absorption edge is obtained at about 290 nm, 280 nm, and 260 nm for ZnS III, ZnS IV, and ZnS V samples, respectively. The gradual shift in absorption edge to the shorter wavelength side (blue shift) indicates increased band gap with reduction in particle size with increasing capping agent.

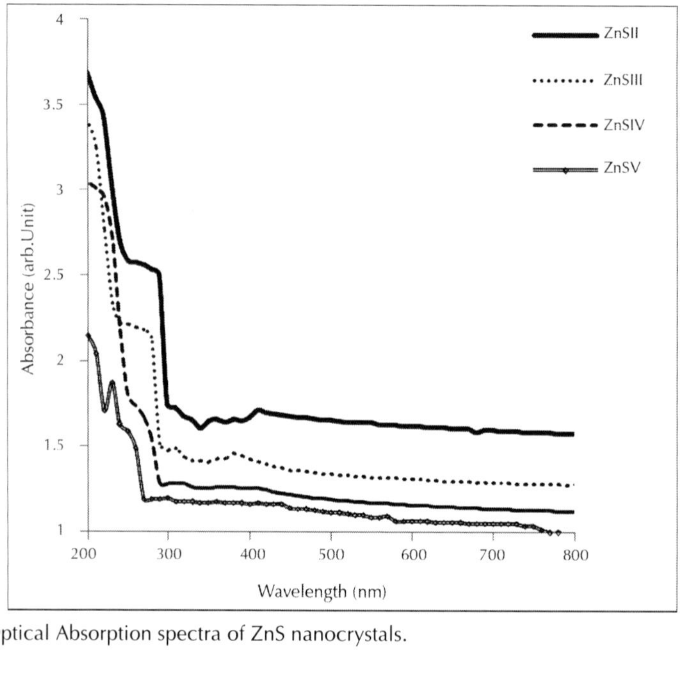

Figure 6. Optical Absorption spectra of ZnS nanocrystals.

The most direct way of extracting the optical band gap is to simply determine the photon energy at which there is a sudden increase in absorption. The optical band gap of the nanocrystalline samples were calculated from the absorption edge and these are given in Table 14.5. The increased band gap with increasing capping agent concentration could be consequence of a "size quantization effect" in the samples as expected; increase in capping agent concentration causes reduction in particle size and therefore increase in the band gap of the samples. The confinement of the electron in the nanocrystal causes the quantization of the energy spectrum in conduction band, which gives rise to a blue shift of the threshold of absorption or absorption peak with decreasing crystalline size (Nanda et al., 1998) The effective optical band gap is given by:-

$$Eg' = Eg + \frac{h^2\pi^2}{2r^2}\left[\frac{1}{m_e^*} + \frac{1}{m_h^*}\right] \tag{6}$$

where E_g' is band gap of nanocrystals, E_g is the band gap of bulk material, r is the radius of cluster, m_e* and m_h* are the effective mass. of electron and hole respectively The above equation indicate the increase in band gap with decreasing size of the nanocrystals. The values of effective mass of electrons and holes for the ZnS are $m_e^* = 0.25$ m_o and $m_h^* = 0.59$ m_o (Brus, 1983), where m_o is the rest mass of electron. Substituting all values in equation (6), we have:

$$r = \frac{1.65x\ 10^{-9}\ m_0}{(E_g'\text{-}E_g)^{1/2}} \tag{7}$$

The crystal size of the ZnS nanocrystal is estimated by the change in the absorption edge using effective mass approximation (EMA) model. The estimated size of ZnS nanocrystal with change in capping agent concentration is given in Table 14.5. It is seen that the particle size is decreased with increasing capping agent concentration. It is observed that no optical absorption occurs at surface states and therefore these do not affect the absorption spectra. Only the widening of the band gap is indicated. Similar results are reported by Kumbhojkar et al. (2000) on mercaptoethanol capped ZnS nanoparticles. They have obtained a broad featureless and small absorption peak at 280 nm. Martinez et al. (Martinez-Caston et al., 2005) have studied absorption spectra of CdS nanoparticles. They have not obtained any peak. Band gap computed from absorption edge has been obtained as 2.57 eV. The estimated particle size is nearly same as obtained from XRD.

Table 14.5. Estimated particle size from absorption edge of ZnS nanocrystals.

Sample name	Concentration of capping agent	Absorption edge wavelength λ (nm)	Energy band gap Eg (eV)	Radius of crystal r (nm)	Crystal size D (nm)
ZnS II	.005 M	300	4.13	2.45	4.9
ZnS III	.01M	290	4.27	2.1	4.2
ZnS IV	.015M	280	4.42	1.90	3.81
ZnS V	.02M	260	4.76	1.55	3.1

Absorption Spectra of ZnS/PVA Nanocomposite Film

For studying optical absorption spectra of ZnS/PVA composite, the film was cast on glass substrate, and a similar glass plate was taken in reference. Figure 14.7 shows the absorption spectra of the ZnS/PVA films with different loading of ZnS. The fundamental absorption, which corresponds to electron excitation from valence band to conduction band, can be used to determine the nature and value of optical band gap. It is noticed that the absorption edge is shifted to shorter wavelength with increasing the loading of ZnS at wt%. The optical band gaps for different samples are listed in Table 14.6.

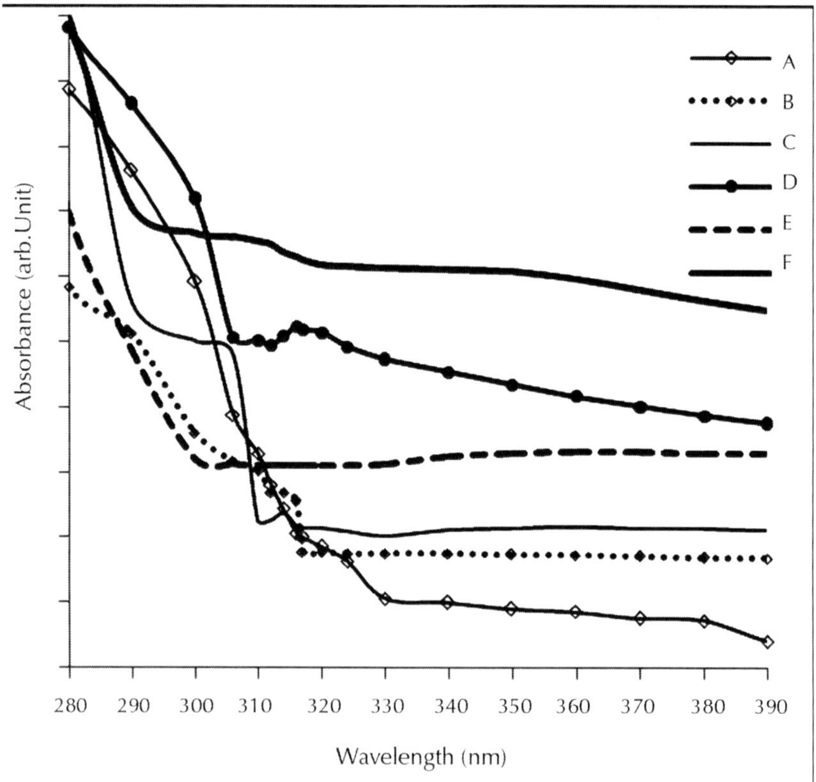

Figure 7. Absorption spectra of ZnS/PVA nanocomposite films with different loading of ZnS.

Rathore et al. have also reported the increase in the band gap with increasing the molar concentration of parent solution of $ZnCl_2$ (Rathore et al., 2008). Here, the wt % of zinc acetate is represented by loading % of ZnS. The energy band gap is increased with higher loading of ZnS in PVA matrix due to formation of smaller crystallite, which is confirmed by the XRD studies also. The particle size is estimated by the effective mass approximation model using equation (7). The estimated particle size is given in Table 14.6, which shows that when the loading (% at wt) of ZnS is increased, the particle size is decreased. The absorbance is also increased with increasing the loading of ZnS in PVA matrix, because of the transparency of the film is reduced with higher loading of ZnS.

Table 14.6. Estimated particle size from absorption edge of ZnS/PVA nanocomposite film.

Sample name	% (at wt)ZnS	Absorption edge wavelength λ (nm)	Energy band gap E_g (eV)	Radius of crystal r (nm)	Crystal size D (nm)
A	2%	327	3.79	4.9	9.8
B	5%	316	3.92	3.3	6.7
C	10%	310	4.0	2.91	5.8
D	20%	305	4.06	2.6	5.3
E	30%	300	4.13	2.4	4.8
F	40%	290	4.27	2.1	4.2

PHOTOLUMINESCENCE (PL)

The PL studies provide information of different energy states available between valence and conduction bands responsible for radiative recombination. In PL, a material gains energy by absorbing photons and promotes electrons from a lower to higher energy level. This may be described as making a transition from the ground state to an excited state of an atom or molecule, or from the valence band to the conduction band (CB) of a semiconductor. In a semiconductor the energy separation, that is the energy difference between the completely filled valence band and the empty CB is of the order of a few electron volts and increases with a decreasing size in nanometer range. Nanocrystals have discrete electron energy levels and surface defects play an important role in their electronic transitions. The smaller particles have higher surface/volume ratio and more surface states. Therefore, they contain more accessible carriers for PL. These indicate that the surface states are very important for the physical properties especially for the optical properties of nanoparticles. Luminescence studies provide information regarding defects states, which take part in radiative de-excitation of the sample. In nanocrystal the defects states may shift or the density may increase which is revealed by the PL studies.

Photoluminescence of ZnS Nanocrystals

A Perkin Elmer LS-55 fluorescence spectrometer in nanophosphor laboratory, Allahabad is used to obtain PL spectra. The PL was excited by different wavelength 225, 275, 300, and, 325 nm. The PL spectra of the nanocrystalline ZnS I, ZnS II, ZnS III, ZnS IV, and ZnS V samples were recorded at room temperature (300 K) in the range from 350 nm to 700 nm. In all the excitation PL show broad emission. When PL was excited by 325 nm, the PL peak at about 450 nm with two side bands at about 420 nm and 480 nm was observed (Figure 14.8). A small hump at 525 nm was also observed, however elongated tail observed suggests the presence of a PL peak at this position. The position of the peak is blue shifted with increasing capping agent concentration but 0.005 M capping agent concentration the poor and red shifted PL peaks are observed in comparison to the bulk ZnS. Hence the PL in this region is due to the presence of impurity levels, or the presence of sulfur vacancies in the lattice, which is also previously, reported by Sooklal et al., (1996) and Bol et al., (1998). It was also observed that with increasing capping agent concentration, intensity of PL peak is increased due to efficient energy transfer from the surface of merceptoethanol to interstitial sites and vacancy centers. Gosh et al., (2006) also have shown increase in PL intensity of capped ZnS nanoparticles in comparison to uncapped ZnS nanoparticles. The broadness in PL spectra for ZnS nanocrystallite can be explained as follows:

The photo-generated charge carriers trapped in shallow states tunnel from one trap to another and recombine with opposite type of charge carrier. The emission from the recombination in shallow traps appears at a lower wavelength than deep traps. The broad emission band represents the superposition of wide distribution of traps distance (Spanhel and Anderson, 1991).

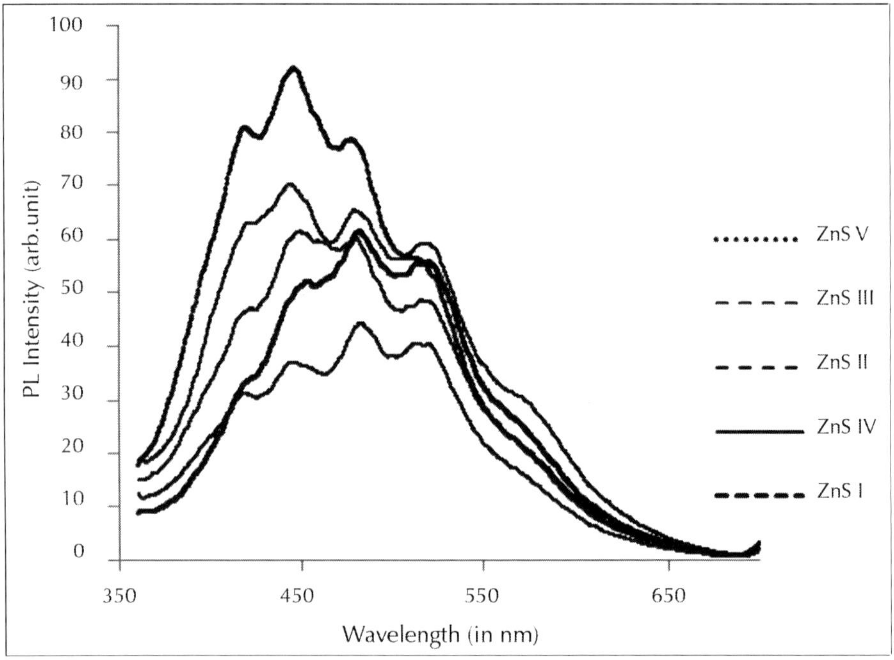

Figure 8. Photoluminescence spectra of ZnS nanocrystal with excitation wavelength 325 nm.

Karar has observed the PL peak at 460 nm in undoped ZnS nanocrystal (Karar et al., 2004). The position of PL peak at 460 nm might be due to native acceptor levels in nanocrystalline ZnS. The origin of this can be native zinc and copper acceptor impurity incorporated during sample preparation (Soo et al., 1994; Yang et al., 1997). The slight shift in peak positions may be due to the nanocrystalline nature of the samples and the related confinement effect.

When PL was excited by 225 nm, the PL spectra of ZnS show a broad emission maximum at about 404 nm (Figure 14.9). This large red shift as observed previously also could be attributed to recombination from surface traps (Becker and Bard, 1983; Yang et al., 1997). Malik et al. have reported the PL of TOPO capped ZnS nanocrystals (Malik et al., 2001). They have found maximum PL at 404 nm. In PL spectra another small peak at 480 nm and small hump at 525 nm was observed, which are also observed in excitation of 325 nm. These may be attributed to sulfur vacancies and impurities. Bhatti et al. (2006) have reported the PL spectra of undoped ZnS nanoparticles at 412 nm and at 433 nm due to Ni impurity in ZnS nanocrystals.

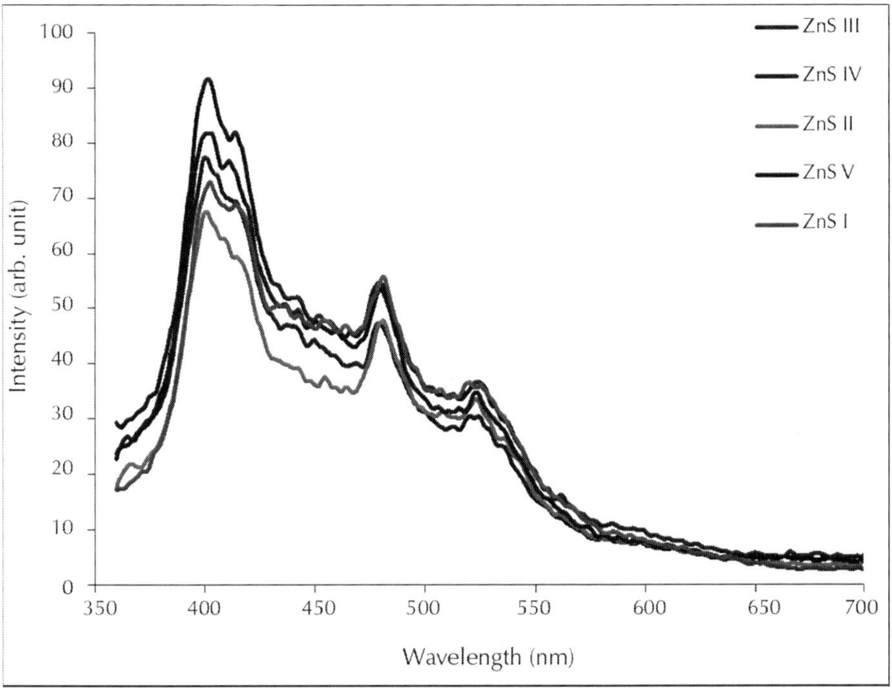

Figure 9. Photoluminescence spectra of ZnS nanocrystal with excitation wavelength 225 nm.

The luminescence mechanism can be understood from the energy level diagram. The energy states within the band gap, in nanocrystals, are produced due to the surface states or Zn^{2+} or S^{2-} ions. Photons of higher energy excite the electron from the valence band or Zn^{2+} levels, which reaches the CB. The excited electrons decay non-radioactively to surface states and then decay radiatively to valence band and emit a photon of lower energy. When the particle size decreases, the valence band edge shifts downwards. Therefore, the emitted photon has comparatively higher energy giving PL peak at shorter wavelength.

In ZnS nanoparticles, a large fraction of total number of atoms resides on the surface. The SH group of mercaptoethanol dissociates and organic group gets attached to

Zn ions. Thus, the organic legends are instrumental in removing Zn dangling orbital from the gap. The SP_3 hybridized orbital of surface S atoms dangle out of the crystal surface. More unsaturated S dangling bonds will be present on the surface. Hence, the legend- terminated surfaces often show deep hole traps. It has been reported in earlier study on colloidal ZnS that the vacancy states lie deeper in the gap than states arising from interstitial atoms (Bruchez et al., 1998). These observations indicate that the PL peak obtained at shorter wavelength with reducing the particle size is due to transition between interstitial states and vacancy states respectively shown in Figure 14.10. It is to be noted that PL peak is obtained in uncapped particles, which indicate intrinsic nature of the peak. Similar results are obtained by Biswas et al., (2006) and Chen et al. (2004). Biswas et al. have reported that PL peak at 400 nm, Chen et al. have compared the results with that of bulk ZnS. They obtained PL emission at 400 nm for ZnS nanoparticles, where as 450 nm for bulk ZnS.

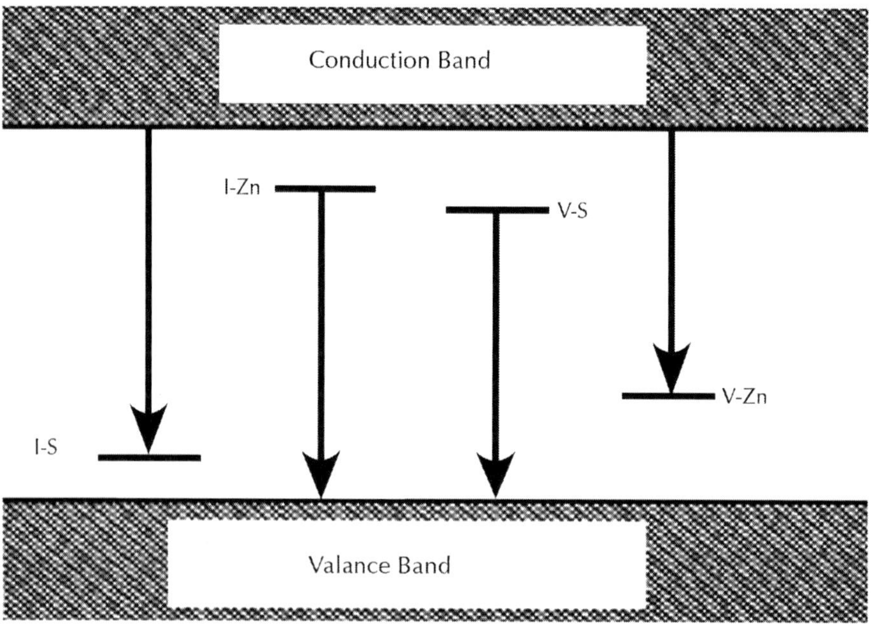

Figure 10. Transitions of ZnS nanocrystals.

Figure 14.11 shows the PL spectra of ZnS V sample with different excitation wavelength. It is seen that for lower excitation wavelength, PL spectrum is broader, which become narrower with higher excitation wavelength and peak position is approximately same in all the excitation wavelength, but in case of 275 nm excitation wavelength, one more peak at 404 nm has been observed. The intensity of PL spectra increases with increasing excitation wavelength. Bhatti et al. (2006) have also found the more intense PL spectra with higher excitation wavelength in case of ZnS nanocrystals.

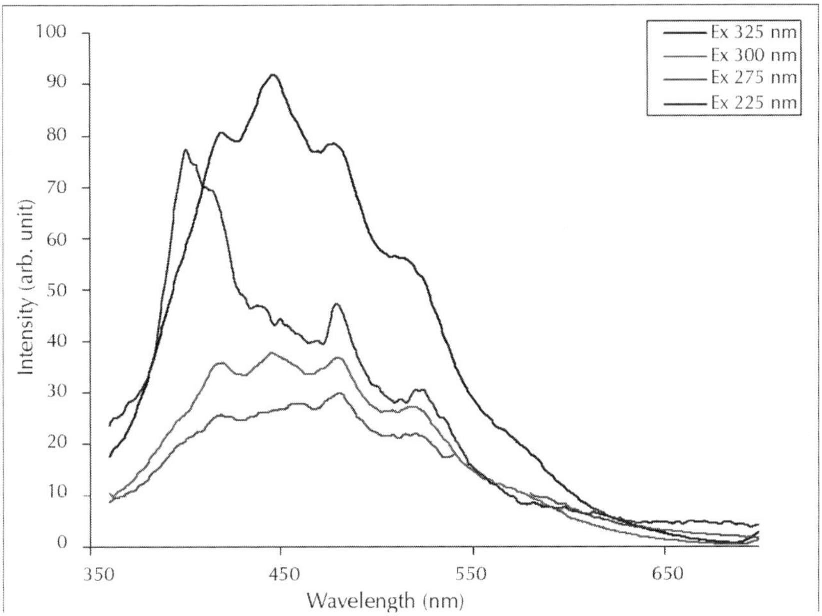

Figure 11. Photoluminescence spectra with different excitation wavelength for ZnS V sample.

Photoluminescence of ZnS/PVA Nanocomposite Film

Photoluminescence of ZnS/PVA Nanocomposite was also studied by Perkin Elmer LS-55 Fluorescence Spectrometer with Xenon lamp in visible region. PL was excited by three different wavelength of light 275, 300, and 325 nm. The PL spectra excited by 275 nm (Figure 14.12) shows a very broad emission peak centered at 430 nm. This emission may be attributed to the recombination via surface localized state (the trap emission) (Comor and Nedeljkovic, 1999; Hebalkar et al., 2001; Wageh et al., 2003). From the Figure 14.12, it is clear that the intensity of PL emission is enhanced with increasing the loading of zinc in ZnS/PVA composite. This results indicate the increasing amount of ZnS in polymer stops the electron – hole recombination on ZnS surface and reduces the non-radiative transition.

When photoluminescence was excited by 300 nm, PL spectra show broadened emission band with multiple peak maxima at 417 nm, 446 nm, and 480 nm (Figure 14.13). The high energy side band is attributed to the presence of large concentration of sulfur vacancy centers (V_s) and interstitial states in the lattice (Dhas et al., 1999). The emission is called self activated emission due to sulfur vacancies contrary to the Zn vacancy related activated emission in bulk ZnS (Manzoor et al., 2003). The first peak around 420 nm has been observed in ZnS/PVA composite film which can be assigned to the recombination of free charge carriers at defect sites, possibly at the surface of ZnS (Manzoor et al., 2003). It is seen that, when loading of ZnS increases, the peaks merge into one another. The PL intensity increases with increasing loading % of ZnS, indicating that the dangling bonds are better passivated with higher loading of ZnS, which also reduces the size of ZnS nanoparticles. In the higher wavelength side no peak has been observed but the tail of the spectrum extending beyond 650 nm.

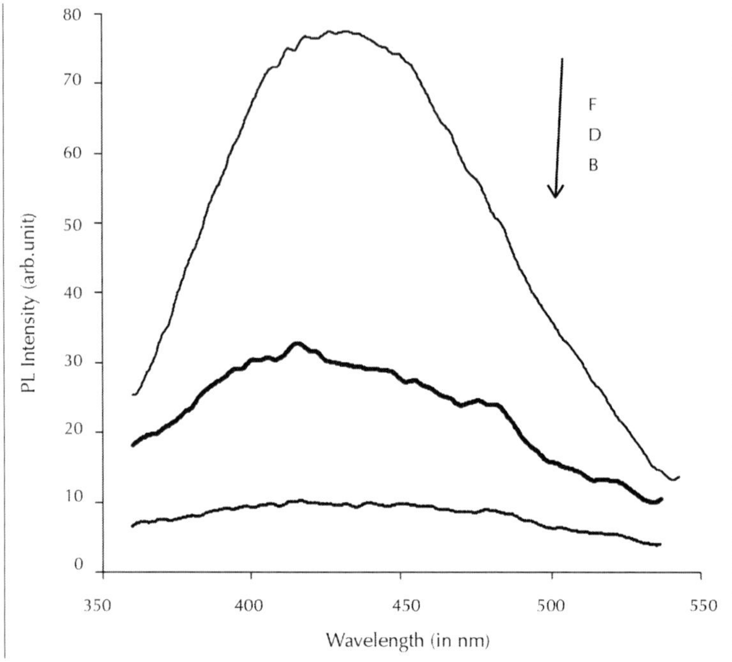

Figure 12. Photoluminescence spectra of ZnS/PVA composite film excited by 275 nm.

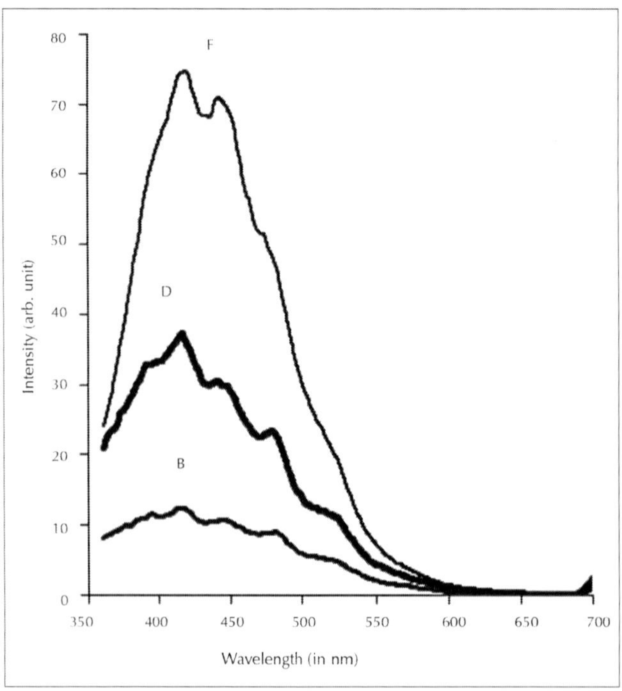

Figure 13. Photoluminescence spectra of ZnS/PVA composite film excited by 300 nm.

The PL spectra of ZnS/PVA composite films excited by 325 nm also show a broad emission (Figure 14.14). In 5% ZnS sample, multiple peaks at 400, 416, and 446 nm has been observed, which are merged and centered around 420 nm with higher loading of ZnS. The PL peak around 400nm is blue shifted with higher loading of ZnS in PVA. This emission could be attributed to recombination from surface traps (Becker and Bard, 1983; Yang et al., 1997). By excitation with 275 and 300 nm light only two peaks around 420 and 445 nm are observed and PL intensity is found to increase with ZnS loading.

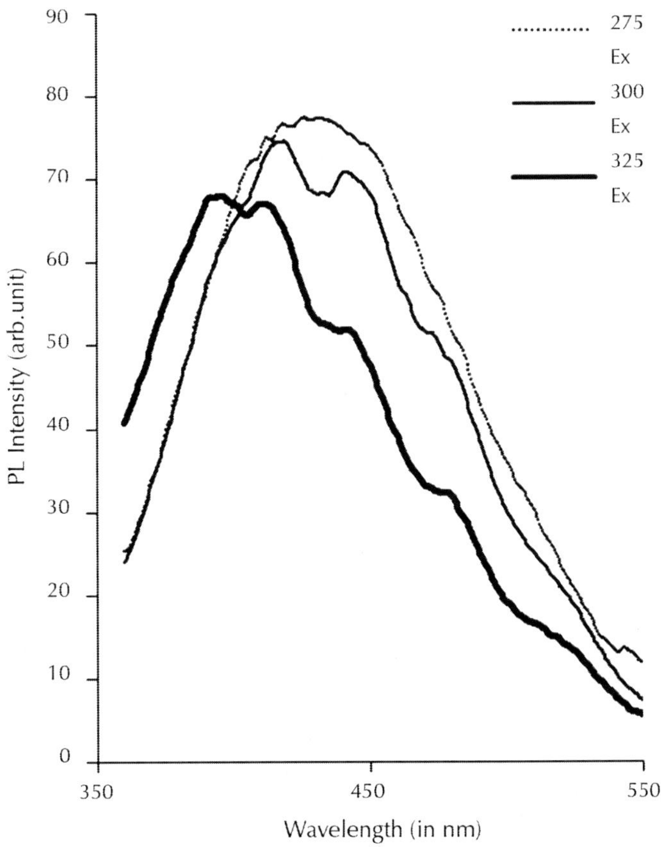

Figure 14. PL spectra of ZnS/PVA composite film excitation by 325 nm.

Figure 14.15 shows the PL spectra of sample F (ZnS 40%) with three different excitation wavelength. It is seen from the figure that, in all the excitation wavelengths, a broad emission band has been obtained. The PL spectrum excited by 275 nm wavelength is more intense and show a single peak around 430 nm, which is split into two peaks at 417 nm and 446 nm with excitation by 300 nm, and in case of excitation by

325 nm, three PL peaks centered at 400, 417, and 446 nm are seen. The intensity of PL emission decreases with excitation wavelength.

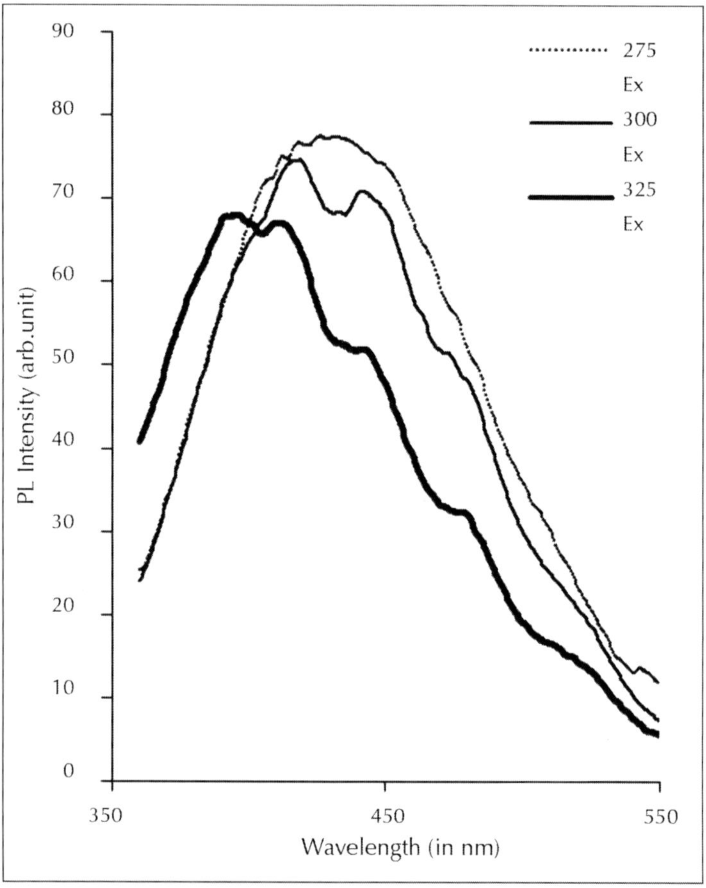

Figure 15. Photoluminescence spectra of sample F with different excitation wavelength.

ELECTROLUMINESCENCE

The phenomenon of emission of light by substances due to application of AC or DC electric field is called electroluminescence (EL). It has been observed in a number of materials in form of powder, thin films, single crystals, p-n junction and metal-semiconductor and metal-insulator-semiconductor structures, and so on. The phenomenon of EL can be considered to be comprised of three sequential processes: (i) excitation, (ii) transfer of energy from site of excitation to that of emission, and (iii) recombination. The EL involves the exciation of luminescence as a result of existence of an applied electric potential difference across the phosphor. The EL properties of nanomaterials and nanocomposites can be significantly controlled by changing the size of the particles.

For study of EL of nanocrystals, the EL cell was prepared by a triple layer structure namely a nanocrystals emission layer, sandwiched between two electrodes; one of the electrodes was transparent for generated light to come out to be measured or used. The transparent electrode has been prepared by depositing thin film of SnO_2 by chemical vapor deposition on clean glass substrate. The cell for EL investigation was fabricated by depositing emission layer of DMSO dispersed ZnS nanocrystals or ZnS/PVA nanocomposite film on SnO_2 coated conducting glass plate. A thin mica sheet having small window of 2x2 mm^2 was fixed over emissive layer of the specimen. An aluminum strip was inserted on it, so as to have a good contact with the emissive layer. Aluminum acts as second electrode.

For EL studies alternating (AC) voltages of various frequencies was applied, by APLAB 2002 sine/square audio frequency generator, to conducting glass plate and aluminum electrode. EL power supply M-EHT-100 AFG (wide band high voltage amplifier) which is supplied by Zesco India, was used for obtaining high voltage. The EL brightness at different voltages and frequencies was measured by a photo multiplier tube (PMT-RCA-931) connected to a picoammeter.

EL spectra were obtained with the help of grating monochromator (HM 104) and photomulitipier tube. The set- up for EL spectra is depicted in Figure 14.16. The EL cell was placed near the entrance slit of grating monochromator. The grating of monochromator was rotated with the help of a drum and then intensity of EL at different wavelength was measured with the help of PMT placed at the exit slit of the monochromator. The output of PMT was obtained using digital picoammeter.

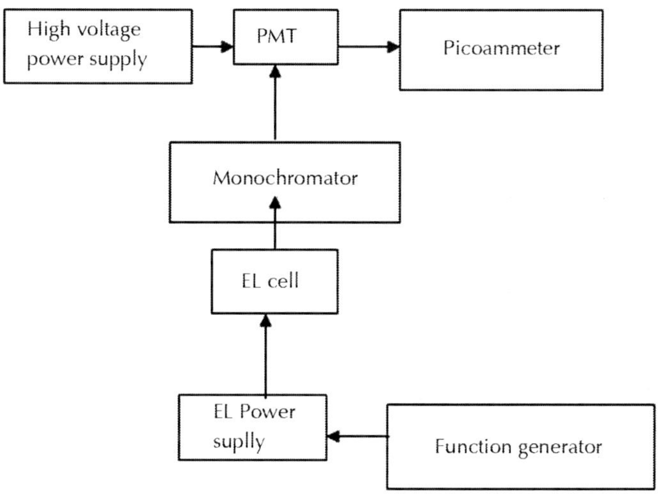

Figure 16. Set- up for Electroluminescence spectra.

The EL studies of ZnS nanocrystals and nanocomposites show violet-blue emission. It is found that the light emission starts at certain threshold voltage and then increases with voltage. The EL brightness has been found to depend on the frequency as well as size of nanoparticles. Various characteristics have been investigated as discussed below.

Brightness-Voltage Characteristics

It is observed that EL starts at a threshold voltage and then increases first slowly and then rapidly with increasing voltage. Figure 14.17 shows EL brightness versus voltage curve for various ZnS nanocrystalline samples at 1KHz frequency. The lower threshold and higher brightness have been observed for smaller nanoparticles prepared from higher concentration of capping agent (.02 M).

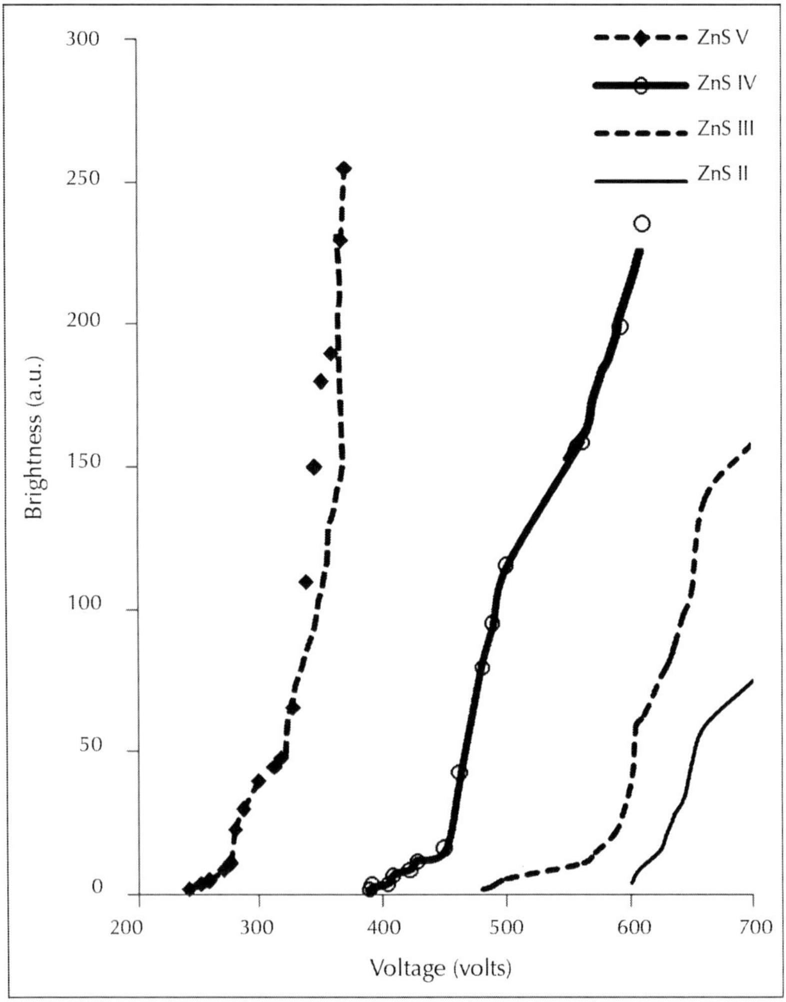

Figure 17. Voltage Brightness characteristic for ZnS nanocrystals at 1KHz.

Similar results are obtained for ZnS/PVA nanocomposite films. The brightness-voltage characteristic curve for composite film is shown in Figure 14.18. As the loading of ZnS in film increased, the operating voltage decreases. It is seen that for higher loading of ZnS, particle size decreases and improved brightness has been observed with lower threshold voltage.

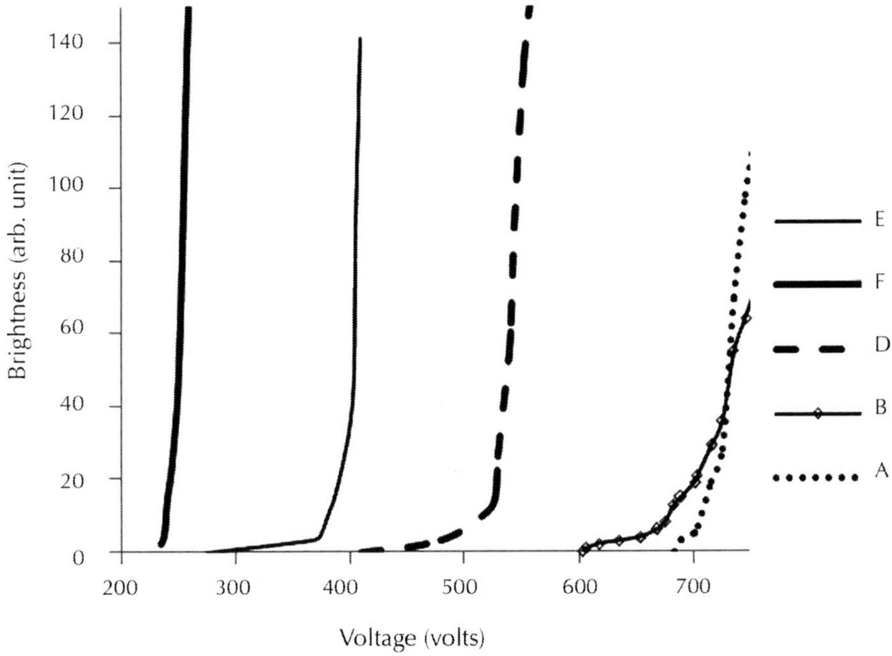

Figure 18. Brightnes-Voltage curve of composite film with different loading of ZnS in PVA.

Effect of Particle size on Threshold Voltage

It can be seen from the Figures 14.17 and 14.18 that the EL emission starts at lower voltages when particle size is reduced either by increasing capping agent in case of nanoparticles or by increasing ZnS loading in case of nanocomposites. Figure 14.19 shows a graph between the concentration of capping agent and threshold voltage. It is clear from the negative slope of the curve that there is linear decrement with increasing the capping agent concentration. From the XRD and absorption study, it is found that the particle size is decreased with increasing capping agent concentration. This reveals that small particles formed with high concentration of capping agent, give higher EL brightness at lower threshold.

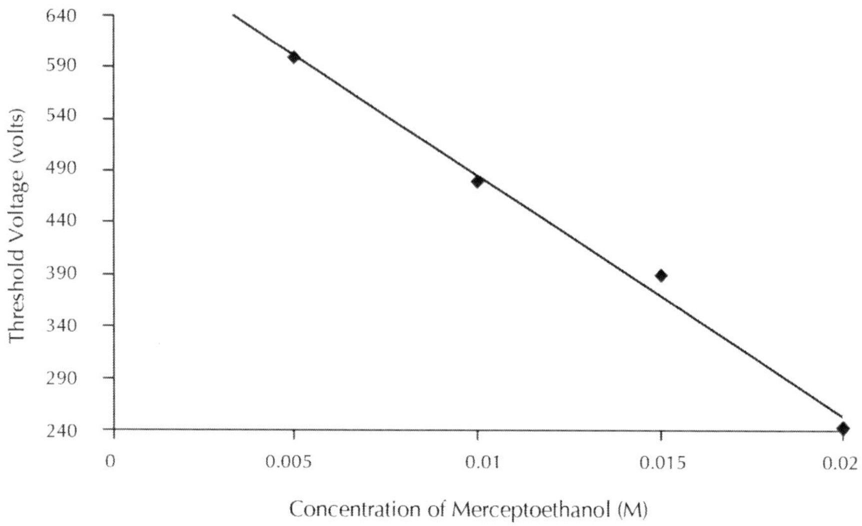

Figure 19. Capping agent concentration versus threshold voltage characteristics.

Figure 14.20 represent a graph between the loading % of ZnS and threshold voltage at different frequncies. It is observed that at the beginning threshold voltage slightly increases with increasing the loading, but at higher % of loading of ZnS in PVA the threshold voltage decreases. From the XRD and absorption studies have shown that smaller ZnS nanoparticles are formed with higher ZnS loading in PVA. Thus here also smaller nanoparticles give EL at lower threshold voltage. It is clear form the figure that the rate of decrease of threshold voltage is more for higher frequencies.

Effect of Frequency on EL Brightness

Figure 14.21 shows the voltage versus EL brightness characteristics of ZnS V sample at different frequencies. It is observed that at higher frequencies, light emission starts at lower threshold voltages, and electroluminescence brightness increases with increasing frequency. The EL brightness at different frequencies is shown in Figure 14.22 for the sample ZnS V at various constant voltages. It is seen that the EL brightness increases rapidly at lower frequencies and attains saturation at higher frequencies.

Figure 14.23 shows the voltage-brightness curve for 40% ZnS /PVA nanocomposite film (Sample F) at different frequencies. As the applied input voltage across EL cell increases, emission starts at a particular threshold voltage and then increases with voltage. The increase in EL intensity is faster for higher frequencies. The frequency-brightness characteristic is depicted in Figure 14.24. It shows that at lower frequencies the brightness increases linearly with frequency and at higher frequencies it attains saturation.

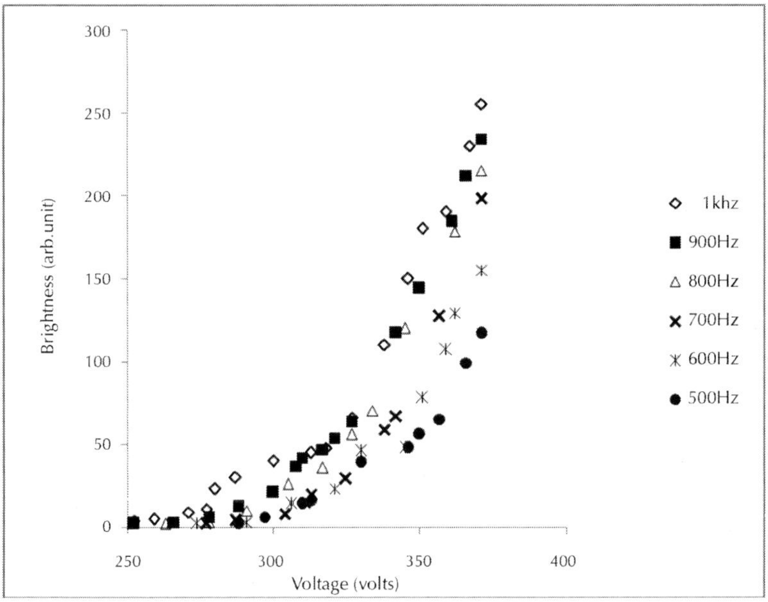

Figure 21. Voltage brightness curve of ZnS V sample a different frequncies.

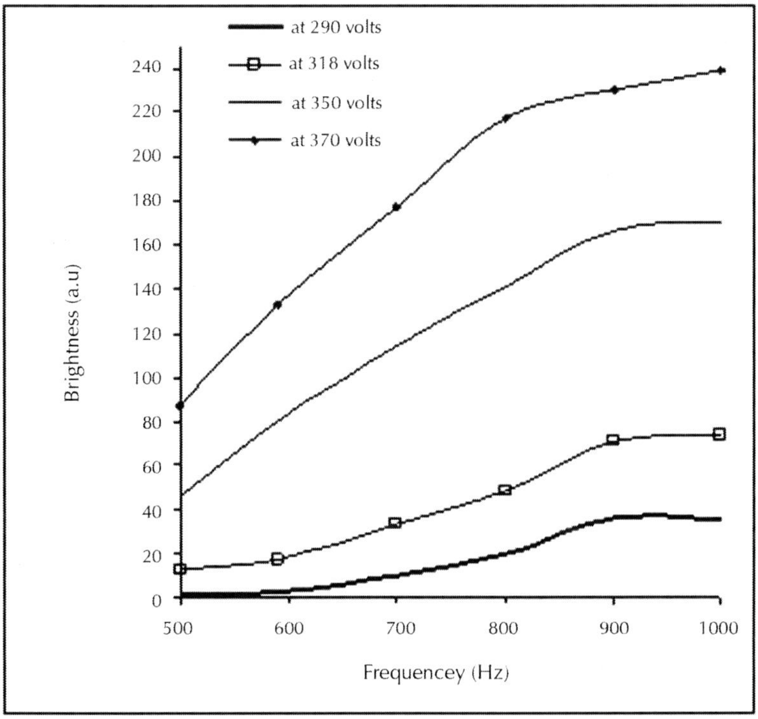

Figure 22. Frequency- Brightness characteristics for Sample ZnS V.

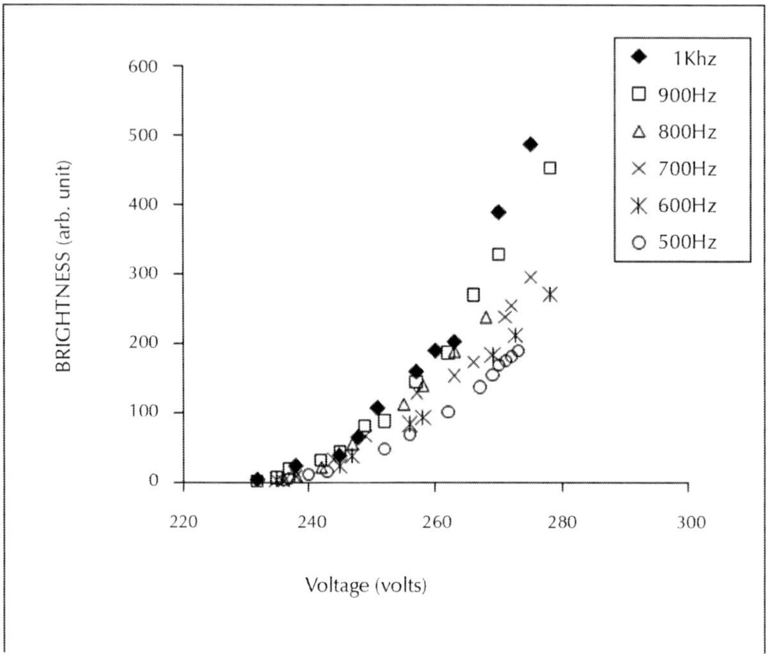

Figure 23. Brightness—Voltage curve for ZnS/ PVA (Sample F) at different frequncies.

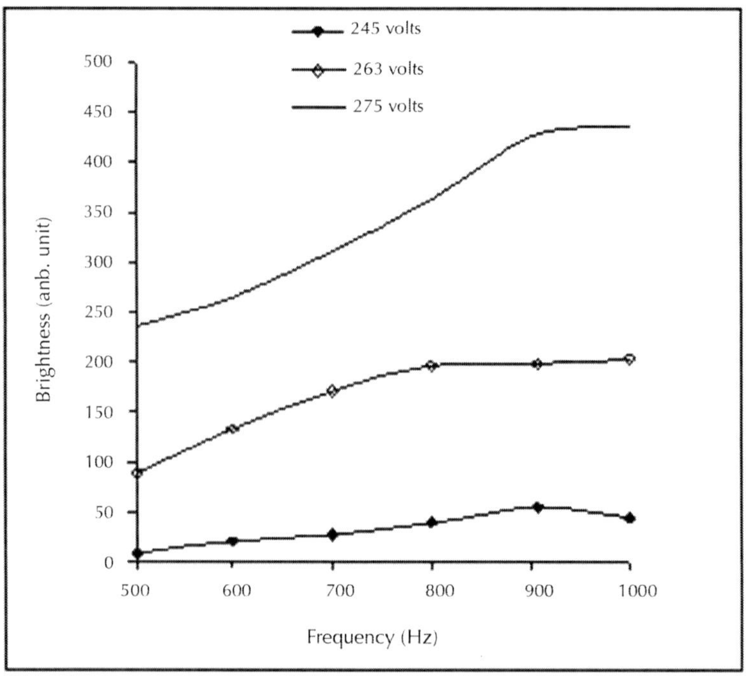

Figure 24. Frequency—Brightness characteristics for ZnS/PVA (sample F).

The nature of frequency dependence of EL brightness can be understood on the basis that the emptying and de-trapping of EL centers more rapidly with increase in frequency. However, at higher frequency the time related to half cycle of the applied A.C. voltage will be comparable to effective time constant for the recombination of electrons. Hence, nonlinearity occurs in the brightness versus frequency plot at higher frequency.

Current-voltage Characteristics

The relation between current through the EL cells and voltage across the cell is found to be linear in case of nanocrystalline ZnS cells as well as for ZnS/PVA cells. The current-voltage characteristics at various frequencies are shown in Figure 14.25 for ZnS III sample. Figure 14.26 depicts similar curves for ZnS/PVA nanocomposite film sample F. The linear relation between current and voltage indicates the ohmic nature. Such an ohmic behavior can be attributed to hopping conductivity of electrons through the nanoparticles in ZnS films. From the I-V characteristics, it is also seen that the slope of the lines increases with frequency in both cases. This indicates lower impedance at higher frequencies.

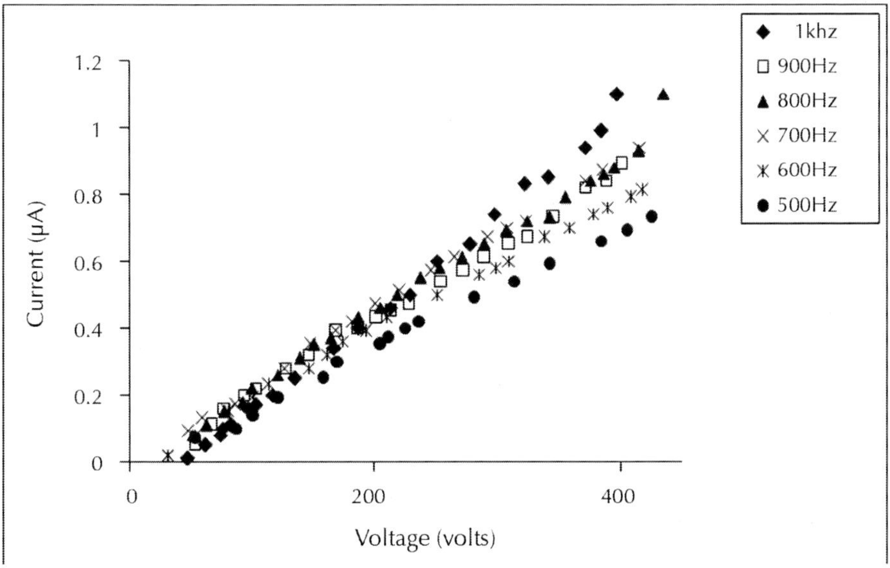

Figure 25. Current—Voltage characteristics for ZnS III Sample.

The current-voltage characteristics of cells with different nanoparticle sizes have shown that small particles prepared with high concentration of capping agent or high ZnS loading in PVA give lower impedance EL cells.

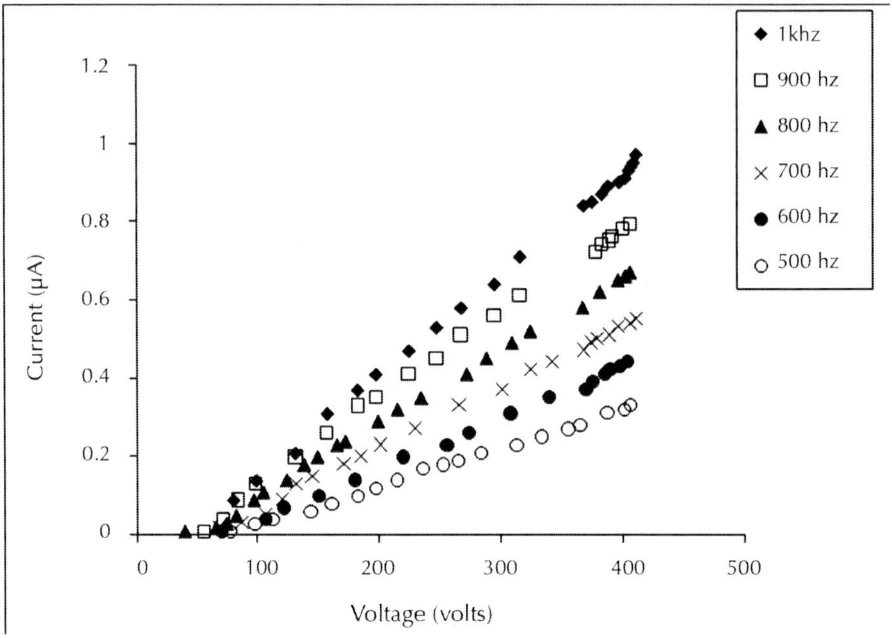

Figure 26. Current—Voltage characterstics forZnS/PVA Sample F (ZnS 40%).

Electroluminescence Spectra

Electroluminescence spectra reveals defect related emission at energies less than the band gap. Figure 14.27 shows the EL spectra of two ZnS/PVA composite film samples with 20% and 40% ZnS loading at 312 volts with fixed 1 KHz frequency. Two peaks are observed at about 425 nm and at about 480 nm. There is little difference between PL and EL spectra; in the PL spectra three peaks are obtained. The peak becomes sharper with higher loading of ZnS. It is speculated that the emission is from the some deep trap luminescence. The results of EL and PL are different due to different excitation mechanism for the two processes (Davis and Williams, 1989). For PL single exciton was excited and then recombines within a very short time at the location where they are formed. However, in EL, charges can be trapped by lower energy states as they are injected and transported in the organic material. While increasing the applied voltage, blue light become visible just below 300 volts. Yang et al have also observed blue light emission with low turn on voltage of single layer ZnS nanoparticles/Polymer composite as an emitter. They have found the hexagonal structure of ZnS/polymer composite film (Yakimov et al., 1995). We have observed high field EL. In this case bending of bands takes place and release of electrons from traps to CB and their subsequent recombination with holes in valence band gives edge emission. On the other hand, if electrons moving in the CB recombine with the holes trapped at defect states, then light emission characteristics of defect centers is observed.

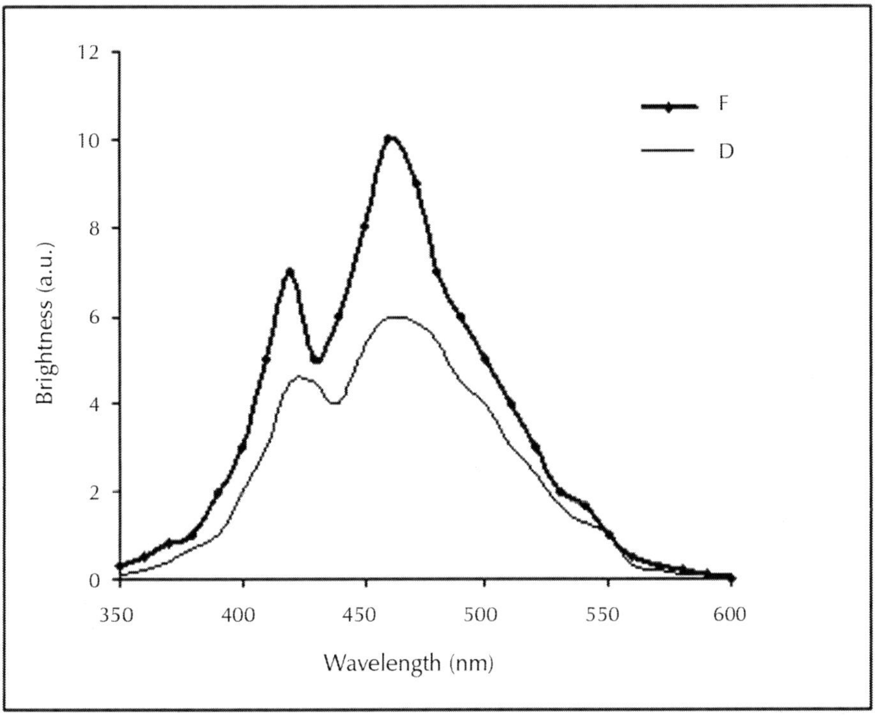

Figure 27. Electroluminescence spectra of ZnS/PVA composite films (voltage 312V).

CONCLUSION

The ZnS nanoparticles and ZnS/PVA nanocomposite films of different crystalline sizes have been synthesized by chemical route. The particle size is controlled by changing the capping agent concentration and loading of ZnS in PVA respectively. The structural and optical characterization of the samples was carried out by XRD and UV-VIS spectroscopy. The XRD study shows formation of ZnS nanocrystals, having size between 2 and 4 nm. Particle size is decreased with increasing capping agent concentration and content of Zn in ZnS/PVA composite films. From the XRD study, ZnS nanocrystals with merceptoethanol capping agent exhibit cubic zinc blend crystal structure, while the ZnS/PVA nanocomposite films show wurtizite crystal structure with hexagonal phase. Optical absorption spectra show the blue shift in absorption edge with increasing capping agent concentration, and higher loading of ZnS in ZnS/PVA nanocomposite film. Absorption increases at lower wavelength and the particle is smaller than Bohr exciton radius and exhibit quantum confinement effect. Photoluminescence (PL) was excited by three different wavelengths. The PL studies of ZnS nanoparticles show the three peaks in violet-blue region of the spectrum (at 450, 480, and 525 nm). The peaks may be due to presences of sulfur vacancies or impurity states in the lattice, which is also previously reported. The intensity of the peaks increases by reducing the particle size with increasing capping agent concentration. The PL spectra of ZnS/PVA nanocomposite film show highly broadened emission band in violet–blue region. The high energy side band is attributed to the presence of sulfur vacancies in the lattice. The intensity of the peaks increases with higher loading of ZnS in PVA matrix. The intensity of PL peaks is greater in case of ZnS naoparticles as compared to that of ZnS/PVA nanocomposite films. This reveals that mercalptoethenol capping agent is more effective to reduce the non-radiative recombinations in ZnS as compared to the polyvinyl alcohol in ZnS/PVA composite film. The electroluminescence studies of ZnS nanoparticles and ZnS/PVA nanocomposite films show lower turn on voltage for smaller nanoparticle with higher brightness. Smaller ZnS nanoparticles have increased oscillator strength, which improves the electron-hole radiative recombination and enhances electroluminescence. There is linear relation between voltage and current. The turn on voltage depend on the size of the particle and the applied frequency. In lower frequency range the brightness of the ZnS nanoparticles and ZnS/ PVA nanocomposite films increases linearly and in higher frequencies it attains saturation. The electroluminescence spectra of ZnS/PVA nanocomposite films show violet-blue light emission with two peaks at 425 and 480 nm The EL intensity is found to increase with higher loading of ZnS in the nanocomposite. The EL emission in ZnS/PVA composite films starts at lower threshold voltage as compared to ZnS nanocrystals of same size. The most interesting fact is that the EL brightness increases very fast and thus quite high brightness can be obtained at comparatively lower voltages. Embedding ZnS nanocrystals in PVA matrix improves the electroluminescent properties. The composite films require much lower voltage for light emission as compared to powder nanocrystalline samples. The hybridization of organic and inorganic semiconductors provide a new approach to construct high performance EL devices with high luminescence efficiency and high carrier density and better transport mechanism.

KEYWORDS

- **Absorption**
- **Electroluminescence.**
- **Nanocomposites**
- **Nanocrystals**
- **Photoluminescence**
- **X ray diffrection**
- **Zinc sulfide**

References

1

Aharoni, S. (1978). *Macromol. Chem.* **179**, 1867-1871.

Akabori, K., Tanaka, K., Kajiyama, T., and Takahara, A. (2003). *Macromolecules* **36**, 4937-4943.

Bershtein, V. A. and Egorov, V. M. (1994). *Differential Scanning Calorimetry of Polymers; Physics, Chemistry, Analysis, Technology*. Ellis Horwood, New York, p. 253.

Boiko, Yu. M. (2000). *Polymer Science* **B42**, 65-67.

Boiko, Yu. M. and Lyngaae-Jørgensen, (2004). *J. Polymer* **45**, 8541-8549.

Boiko, Yu. M. and Prud'homme, R. E. (1997). *Macromolecules* **30**, 3708-3710.

Brown, H. R. and Russell, T. P. (1996). *Macromolecules* **29**, 798-800.

Composto, R. J., Kramer, E. J., and White, D. M. (1992). *Macromolecules* **25**, 4167-4174.

Dalnoki-Veress, K., Forrest, J. A., Murray, C., Gigault, C., and Dutcher, J. R. (2001). *Phys. Rev. E.* **63**, 031801.

De Gennes, P. G. (1971). *J. Chem. Phys.* **55**, 572-579.

Fischer, H. (2002). *Macromolecules* **35**, 3592-3595.

Fu, J., Li, B., and Han, Y. (2005). *J. Chem. Phys.* **123**, 064713.

Green, P. F. and Kramer, E. J. (1986). *Macromolecules* **19**, 1108-1114.

Guérin, G., Mauger, F., and Prud'homme, R. E. (2003). *Polymer* **44**, 7477-7484.

Hwang, E. J. (1997). *Korean J. Rheol* **9**, 89-96.

Hyun, J., Aspnes, D. E., and Cuomo, J. J. (2001). *Macromolecules* **34**, 2395-2397.

Jou, J. (1986). *PhD Thesis*. The University of Michigan. Kajiyama, T., Tanaka, K., and Takahara, A. (1995). *Macromolecules* **28**, 3482-3484.

Kawaguchi, D., Tanaka, K., Kajiyama, T., Takahara, A., and Tasaka, S. (1998). *Macromolecules* **31**, 5150-5151.

Kawaguchi, D., Tanaka, K., Kajiyama, T., Takahara, A., and Tasaka, S. (2003). *Macromolecules* **36**, 1235-1240.

Kawana, S. and Jones, R. A. L. (2001). *Phys. Rev. E.* **63**, 021501.

Keddie, J. L., Jones, R. A. L., and Cory, R. A. (1994a). *Europhys. Lett.* **27**, 59-64.

Keddie, J. L.; Jones, R. A. L., and Cory, R. A. (1994b). *Faraday Discuss* **98**, 219-230.

Kim, J. H. and Wool, R. P. (1983). *Macromolecules* **16**, 1115-1120.

Kline, D. B. and Wool, R. P. (1988). *Polym. Eng. Sci.* **28**, 52-57.

Linghu, X., Zhao, J., Song, R., and Fan, Q. R. (2000). *Chinese Chem. Lett.* **11**, 925-928.

Mansfield, K. F. and Theodorou, D. N. (1991). *Macromolecules* **24**, 6283-6294.

Mattsson, J., Forrest, J. A., and Börjesson, L. (2000). *Phys. Rev. E.* **62**, 5187-5200.

Mayes, A. M. (1994). *Macromolecules* **27**, 3114-3115.

Prucker, O., Christian, S., Bock, H., Rühe, J., Frank, C. W., and Knoll, W. (1998). *Macromol Chem. Phys.* **199**, 1435-1444.

Satomi, N., Takahara, A., and Kajiyama, T. (1999). *Macromolecules* **32**, 4474-4476.

Sharp, J. S. and Forrest, J. A. (2003). *Phys. Rev. Lett.* **91**, 235701.

Van Krevelen, D. W. (1997). *Properties of Polymers*, 3rd ed. Elsevier, Amsterdam.

Whitlow, S. J. and Wool, R. P. (1991). *Macromolecules* **24**, 5926-5938.

Willett, J. L. and Wool, R. P. (1993). *Macromolecules* **26**, 5336-5349.

Wool, R. P. (1995). *Polymer Interfaces: Structure and Strength.* Hanser Press, Munich, p. 494.

Wool, R. P. and O'Connor, K. M. (1981). *J. Appl. Phys.* **52**, 5953-5963.

Zhang, X., Tasaka, S., and Inagaki, N. (2000). *J. Polym. Sci. Part B Polym. Phys*. **38**, 654-658.

2

Aly, A. S., Ali, A. M., and Hawaary, S. E. (1999). *J. Text Assoc.* **60**(1), 25–29.

Burkanudeen, A. and Karunakaran, M. (2002). *Orient. J. Chem.* **18**, 65–68.

Burkanudeen, A. and Karunakaran, M. (2003). *Orient. J. Chem.* **19**, 225.

DeGeiso, R. C., Donaruma, L. G., and Tomic, E. A. (1962). *Anal. Chem.* **34**, 845.

Feng, D., Aldrich, C., and Tan, H. (2000). *Miner. Eng.* **13**, 623–642.

Gregor, H. P., Littinger, L. B., and Loebl, E. M. (1962). *J. Physical Chem.* **59**, 34.

Gupta, R. H., Zade, A. B., and Gurnule, W. B. (2008). *J. Appl. Polym.* **109**, 3315–3320.

Gurnule, W. B., Juneja, H. D., Paliwal, L. J., and Kharat, R. B. (2002). *React. Funct. Polym.* **50**, 95.

Gurnule, W. B., Rahangdale, P. K., Paliwal, L. J., and Kharat, R. B. (2003a). *Synth. React. Inorg. Met. Org. Chem.* **33**, 1187.

Gurnule, W. B., Rahangdale, P. K., Paliwal, L. J., and Kharat, R. B. (2003b). *React. Funct. Polym.* **55**, 255–268.

Jadhao, M. M., Paliwal, L. J., and Bhave, N. S. (2005a). *J. App. Polym. Sci.* **96**, 1606–1610.

Jadhao, M. M., Paliwal, L. J., and Bhave, N. S. (2005b). *Ind. J. Chem.* **44**, 542.

Lokhande, R. S. and Singare, P. U. (2007). *Radiochim. Acta.* **95**, 173.

Lokhande, R. S. and Singare, P. U. (2008). *J. Porous. Mater*. **15**, 253.

Lokhande, R. S., Singare, P. U., and Dole, M. H. (2006). *J. Nucl. Radiochim. Sci.* **7**, 29.

Lokhande, R. S., Singare, P. U., and Dole, M. H. (2007). *Radiochemistry* **49**, 519.

Lokhande, R. S., Singare, P. U., and Karthikeyan (2007). *Russ. J. Phys. Chem. A.* **81**, 1768.

Lokhande, R. S., Singare, P. U., and Kolte, A. R. (2007). *Radiochim. Acta.* **95**, 595.

Lokhande, R. S., Singare, P. U., and Patil, A. B. (2007). *Radiochim. Acta.* **95**, 173.

Manavalan, R. and Patel, M. M. (1983). *Micromole. Chem.* **184**, 717–723.

Michel, P. E. P., Barbe, J. M., Juneja, H. D., and Paliwal, L. (2007). *Europ. Polym. J.* **42**, 4995–5000.

Parrish, J. R. (1955). *Chem. Ind. London,* 386.

Parrish, J. R. (1956). *Chem. Ind. London,* 137.

Patel, M. M., Kapadia, M. A., Patel, D. P., and Joshi, J. D. (2007). *React. Funct. Polym.* **67**, 746–757.

Patel, S. A., Shah, B. S., Patel, R. M., and Patel, P. M. (2004). *Iran Polym. J.* **13**, 445–453.

Rivas, B. L., Pereiva, E. D., Gallegos, P., and Geckeler, K. E. (2002). *Polym. Adv. Technol.* **13**, 1000.

Roy, P. K., Rawat, A. S., and Rai, P. K. (2004). *J. Appl. Polym. Sci.* **94**, 1771.

Shah, A. B., Shah, A. V., and Bhandari, B. N. (2001). *Asian J. Chem.* **13**, 1305–1308.

Shah, A. B., Shah, A. V., and Bhandari, B. N. (2008). *J. Iran Chem. Soc.* **5**, 25–28.

Shah, A. B., Shah, A. V., and Shah, M. P. (2006). *Iran Polym. J.* **15**, 809–819.

Singru, R. N., Zade, A. B., and Gurnule, W. B. (2008). *J. App. Polym. Sci.* **109**, 859–868.

Spedding, F., Fulmer, E., Powell, J., Butler, T., and Yaffe, S. (1951). *J. Am. Chem. Soc.* **73**, 4840–4847.

Spedding, F., Fulmer, E., Powell, J., Butler, T., and Yaffe, S. (1952). *Chem. Eng. News.* **30**, 1200–1201.

Suzuki, E. (2002). *J. Microscopy*. **208**, 153–157.

Von Lillin, H. (1954). *Angew. Chem.* **66**, 649.

3

Abou El Naga, H. H., Abd El Azeim, M. W., and Mazin, A. S. (1998). The effect of the aromaticity of base stocks on the viscometric properties of multigrade oils. *Lub. Sci.* **10**(4), 343–363.

Alexander, D. L., Kapuscinski, M. M., and Laffin, M. V. (1989). Shear stability index of VI improvers; It is dependency on method of determination. *Lub. Engin.* **45**(12), 801–806.

Bartz, W. J. (1999). Influence of viscosity index improver, molecular weight, and base oil on thickening tendency, shear stability, and evaporation losses of multigrade oils. *Lub. Sci.* **12**(1), 215–237.

Coutinho, F. M. B. and Teixeira, S. C. S. (1993). Polymer used as viscosity index improver: A comparative study. *Poly. Test* **12**, 415–422.

Hassanean, M. H. M., Bartz, W. J., Abou, H. H., and Naga, El. (1994). A study of viscometric properties of multigrade lubricants. *Lub. Sci.* **6**(2), 149–159.

Jain, P. K., Hemalatha, C. Y., Sundarrajan, S., Girija, N., and Sarma, A. S. (2000). *Effect of base oil characterization on shear stability of viscosity modifiers*. 2nd International Symposium on Fuels and Lubricants, New Delhi.

Singh, H. and Gulati, I. B. (1987). Influence of base oil refining on the performance of viscosity index improvers. *Wear.* **118**, 33–56.

Wardle, R. W. M., Coy, R. C., Cann, P. M., and Spikes, H. A. (1990). An 'In lubro' study of viscosity index improvers in end contacts. *Lub. Sci.* **3**(1), 45–61.

Warren, M. J., Asfour, A. F. A., and Gao, J. Z. (2005). Viscometric behaviour of viscosity index improvers in lubricant base oil over a wide temperature range: II. Polyalphaolefin Synthetic Base oil. *J. Synth. Lub.* **22**, 249–258.

4

Brunel, R., Marestin, C., Martin, V., Mercier, R., and Schiets, F. (2008). *High Perform. Polym.* **20**, 185.

Dominguez, D. D., Jones, H. N., and Keller, T. M. (2004). *Polym. Comp.* **25**, 5.

Earle, M. J. and Seddon, K. R. (2000). *Pure. Appl. Chem.* **7**, 1391.

Giribabu, L., Vijay Kumar, Ch., Surendar, A., Gopal Reddy, V., Chandrasekharan, M., and Yella Reddy, P. (2007). *Synthetic Commun.* **37**, 4141.

Guzman-Lucero, D., Likhanova, N. V., Hopfl, H., Guzmán, J., and Likhatchev, D. (2006). *ARKIVOC Part* X, 7.

Hoffmann, J., Nüchter, M., Ondruschka, B. and Wasserscheid, *P. Green Chem.* **5**, 296.

Holbrey, J. D. and Seddon, K. R. (1999). *Clean Products and Procs.* **1**, 223.

Hou, Y. and Baltus, R. E. (2007). *Ind. Eng. Chem. Res.* **46**, 8166.

Kappe, C. O. (2004). *Angew. Chem. Int. Ed.* **43**, 6250.

Keller, T. M. (1987). *J. Polym. Sci. Part A. Polym Chem.* **25**, 2569.

Keller, T. M. (1988). *Chemtech* **18**, 635.

Keller, T. M. (1993). *Polymer* **34**, 952.

Keller, T. M., and Dominguez, D. D. (2005). *Polymer* **46**, 4614.

Keller, T. M. and Price, T. R. (1984). *Polymer Commun.* **25**, 42.

Keller, T. M. and Price, T. R. (1985). *Polymer Commun.* **26**, 48.

Kubisa, P. (2004). *Prog. Polym. Sci.* **29**, 3.

Kumar, A. and Pawar, S. S. (2007). *J. Org. Chem.* **72**, 8111.

Kumar, D., Razdan, U., and Gupta, A. D. (1993). *Journal of Polym. Sci. Part A: Polym. Chem.* **31**, 2319.

Laskoski, M., Dominguez, D. D., and Keller, T. M. (2005). *J. Polym. Sci. Part A; Polym. Chem.* **43**, 4136.

Laskoski, M., Dominguez, D. D., and Keller, T. M. (2007). *Polymer* **48**, 6234.

Lidstrom, P., Tierney, J., Wathey, B., and Westman, J. (2001). *Tetrahedron* **57**, 9225.

Perreux, L. and Loupy, A. (2001). *Tetrahedron* **57**, 9199.

Sastri, S. B., Armistead, J. P., and Keller, T. M. (1997). *Polym. Comp.* **18**, 48.

Sastri, S. B. and Keller, T. M. (1998). *J. Polym. Sci: Part A; Polym. Chem.* **36**, 18.

Sastri, S. B. and Keller, T. M. (1999). *J. Polym. Sci: Part A; Polym. Chem.* **37**, 2105

Selvakumar, P. and Sarojadevi, M. (2009). *Macromol. Symp.* **277**, 190.

Yeganeh, H., Tamami, B., and Ghazi, I. (2004). *Eur. Polym. J.* **40**, 2059.

Westaway, K. C. and Gedye, R. J. (1995). *Microw. Power and Electromagn. Energy* **30**, 219.

Wiesbrock, F., Hoogenboom, R., and Schubert, U. S. (2004). *Macromol. Chem. Rapid Commun.* **25**, 1739.

5

Abbernt, S., Plestil, J., Hlavata, D., Lindgren, J., Tegenfeldt, J., and Wendsjo, A. (2001). *Polymer* **42**, 1407-1416.

Appetecchi, G. B., Croce, F., Perci, L., Ronci, F., and Scrosati, B. (2000a). *J. Electrochem. Soc.* **147**, 4448-4452.

Appetecchi, G. B., Croce, F., Perci, L. Ronci, F., and Scrosati, B. (2000b). *J. Electrochem. Soc.* **45**, 1481-1490.

Appetecchi, G. B. Hassoun, G. B., Scrosati, B., Croce, F., Cassel, F., and Salomon, M. (2003). *J. Power Sources* **124**, 246-253.

Armand, M. B., Chabagno, J. M., and Duclot, M. (1997). In *Fast ion transport in solids*. P. Vashista, J. N. Mundy, and G. K. Shenoy (Eds.). Elsevier, Amsterdam, p. 131.

Capuglia, C., Mustarelli, P., Quartarone, E., Tomasi, C., and Magistris, A. (1999). *Solid State Ionics* **118**, 73-79.

Capuoano, F., Croce, F., and Scrosati, B. (1991). *J. Electrochem. Soc.* **138**, 1918-1922.

Chiang, C., Chu, Y. P. P., and Reddy, M. J. (2004). *Solid State Ionics* **175**, 631-635.

Chu, P. P, Reddy, M. J., and Kao, H. M. (2003). *Solid State Ionics* **156**, 141-153.

Chusid, O., Gofer, O, Aurbach, Y., Wattanable, D., Momma, M., and Osaka, T. (2001). *J. Power Sources* **97-98**, 632-636.

Croce, F., Appetecchi, G. B., Perci, L., and Scrosati, B. (1998). *Nature* **394**, 456-458.

Fan, L. Z. and Maier, J. (2006). *Electrochem. Commun.* **8**, 1753-1756.

Focher, B., Naggi, A., Torri, G., Cosari, A., and Terbjerich, M. (1992). *Carbohydr. Polym.* **17**, 97-102.

Gorecki, W., Andreani, R., Bertheir, C., Malli, M., and Roose, J. (1986). *Solid State Ionics* **18-19**, 295-299.

Gray, F. M. (1991). *Solid polymer electrolytes-fundamentals and technological applications*. VCH, New York.

Gray, F. M. (1997). Polymer electrolytes, RSC materials monographs. The Royal Society of Chemistry, Cambridge.

Gray, F. M., McCollum, J., and Vincent, R. C. A. (1986). *Solid State Ionics* **18-19**, 282-286.

Huang, B., Wang, Z., Chen, L., Zue, R., and Wang, F. (1996). Solid State Ionics 91, 279-284.

Itoh, T., Miyamura, Y., Iohikawa, Y., Uno, T., Kubo, M., and Yamamoto, O. (2003). *J. Power Sources* **119–121**, 403-408.

Kelly, I. E., Owen, J. R., and Steele, B. C. H. (1985). *J. Power Sources* **14**, 13-21.

Kumar, B. and Scanlon, L. G. (1994). *J. Power Sources* **52**, 261-268.

Li, G., Li, Z., Zhang, P., Zhang, H., and Wu, Y. (2008). *Pure and Appl. Chem.* **80**, 2553-2563.

Manuel Stephan, A. (2006). *Eur. Polym. J.* **42**, 21-42.

Manuel Stephan, A. and Nahm, K. S. (2006). *Polymer* **47**, 5952-5964.

Manuel Stephan, A., Kumar, T. P., Nathan, M. A. K., and Angulakshmi, N. (2009). *J. Phys. Chem. B* **113**(7), 1963-1971.

McCallum, J. R. and Vincent, C. A. (1987). *Polymer electrolytes reviews-I*. Elsevier, London.

Pearson, F. G., Marchessault, R. H., and Liang, C. Y., (1960). *J. Polym. Sci.* **43**, 101-116.

Rajendran, S., Mahendran, O., and Kannan, R., (2002). *Mater. Chem. Phys.* **74**, 52-57.

Ribeiro, R., Silva, G. G., and Mohallen, N. D. S. (2001). *Electrochim. Acta.* **46**, 1679-1686.

Robitailla, C. D. and Fauteaux, D. (1986). *J. Electrochem. Soc.* **133**, 315-325.

Saikia, D. and Kumar, A. (2004). *Electrochimica. Acta.* **49**, 2581-2589.

Schechter, A. and Aurbach, D. (1999). *Langmuir* **15**, 3334-3342.

Scrosati, B. (1993). *Applications of electroactive polymers*. Chapman Hall, London.

Scrosati, B., Croce, F., Perci, L., Serrino-Fiory, F., Plichta, E., and Hendrickson, M. A. (2001). *Electrochim. Acta.* **46**, 2457-2461.

Shin, J. H., Alessandrini, F., and Passerini, S. (2005). *J. Electrochem. Soc.* **152**, A283-A288.

Shodai, T., Owens, B. B., Ostsuka, M., and Yamaki. J. (1994). *J. Electrochem. Soc.* **141**, 2978-2981.

Song, J. Y. and Wang, H. (1999). *J. Power Sources* **77**, 135-146.

Thomas, S. and Thomas, Sabu (2009). *J. Phys. Chem. C.* **113**, 97-104.

Wang, L., Yang, W., Li, X., and Evans, D. G. (2010). *Electrochem. Solid State Lett.* **13**, A7-A10.

Wang, X. J., Kang, J. J, Wu, Y. P., and Fang, S. B. (2003). *Electrochem. Commn.* **5**, 1025-1029.

Weston, J. E. and Steele, B. C. H. (1982). *Solid State Ionics* **7**, 75-79.

Wieczorek, W. (1992). *Mater. Sci. Eng. B* **15**, 108-114.

Wieczorek, W., Steven, J. R., and Florjanczyk, Z. (1996). *Solid State Ionics* **85**, 67-72.

6

Azemi, S. and Bijarimi, M. P. (2004). "Cyclic crack growth measurement using split-tear test piece." *J. Rubb. Res.*

Azemi, S. and Thomas, A. G. (1988). "Tear Behavior of Carbon Black-Filled Rubbers". L. L Amin and K.T. Lau (Eds.). Proceedings International Rubber Technology Conference, Penang, Malaysia 147. Kuala Lumpur: Rubber Research Institute of Malaysia.

Greensmith, H. W. (1956). "Rupture of rubber. IV. Tear properties of vulcanizates containing carbon black." *J. Polym. Sci.* **21**, 175.

Hamed, G. R. and Al-Sheneper, A. A. (2003). "Effect of carbon black concentration on cut growth in NR vulcanizates". *Rubb. Chem. Technol.* **76**. 436.

Lake, G. J. (1972). "Mechanical fatigue of Rubber". *Rubb. Chem. Technol.* **45**, 309.

Lake, G. J. and Lindley, P. B. (October, 1964a) "Cut growth and fatigue of rubber." *Rubb. J.*, 33.

Lake, G. J. and Lindley P. B. (1964b). "Cut growth and fatigue of rubbers II. Experiments on non-crystallizing rubbers." *J. Appl. Polym. Sci.* **8**, 707–721.

Lindley, P. B. (1973). "Relation between hysteresis and dynamic crack growth resistance of NR." *Int. Journal Fracture* **9**, 449.

Payne, A. R. and Whittaker, R. E. (1971). *Rubb. Chem. Technol.* **44**, 440–478.

Rivlin, R. S and Thomas, A. G. (1953). "Rupture of rubber 1. Characteristic energy for tearing". *J. Polym. Sci.* **10**, 291–318.

Thomas, A. G. (1958). "Rupture of rubber Part 5."Cut growth in natural rubber vulcanizates. *J. Polym. Sci.* **31**, 467.

Thomas, A. G. (1960). "Rupture of rubber. Part 6: Further experiments on the tear criterion". *J. Appl. Polym. Sci.* **3**, 168.

Yoshihide Fukahori. (2003). "The mechanics and mechanism of the carbon black reinforcement of elastomers." *Rubb. Chem. Technol.* **76**, 548.

7

Agrawal, R. C. (2008). In *Solid State Ionics: New materials for pollution free energy devices.* B. V. R. Choudhary et. al. (Eds.). Macmillan Press. Coimbatore, India.

Ahmad, S. and Agnihotry, S. A. (2009). *Current Applied Science* **9**, 108.

Akaram, Md., Javed, A., and Rizvi, T. Z. (2005). *Turk. J. Phys.* **29**, 355.

Appetecchi, G. B., Croce, F., Moyroud, E., and Scrosati, B. (1995). *J. Appl. Electrochem.* **25**, 987.

Bhargav, P. B., Mohan, V. M., Sharma, A. K., and Rao, V. V. R. N. (2007). *Ionics* **13**, 173.

Bohnke, O., Rousselot, C., Gillet, P. A., and Truche, C. (1992). *J. Electrochem. Soc.* **139**, 1862.

Chandra, S. and Singh, N. (1983). *J. Phys. C. Solid State Physics* **16**, 3081.

Chen, W., Xu, Q., and Yuan, R. Z. (2000). *Material Science and Engineering* B **77**, 15.

Christie, A. M., Lilley, S. J., Staunton, E., Andreev, Y. G., and Bruce, P. G. (2005). Increasing the conductivity of crystalline polymer electrolytes. *Nature* **433**, 50.

Dyer, John R. (1991). *Applications of absorption spectroscopy of organic compounds*. Prentice Hall of Pvt. Ltd., New Delhi.

Gray, F. M. (1991). *Solid Polymer Electrolytes: Fundamental and Technological Applications*. VCH Publications, New York.

Howe, A. T. and Shilton, M. G. (1979). *J. Solid State Chem.* **28**, 345.

Kang, S. W., Kim, J. H., Char, K., Won, J., and Kang, Y. S. (2006). *J. Membrane Sci.* **285**, 102.

Mohamad, A. A. and Arof, A. K. (2006). *Ionics* **12**, 263.

Pandey, K., Dwivedi, M. M., Tripathi, M., Singh, M., and Agrawal, S. L. (2008). *Ionics* **14**, 515.

Prajapati, G. K., Roshan R., and Gupta, P. N. (October, 2010). *J. of Physics and Chemistry of Solid*s (in press).

Scrosati, B. (1993). *Application of Electroactive Polymers*. Chapman & Hall publications, London.

Sekhon, S. S., Pradeep, K. V., and Anihotri, S. A. (1998). In *Solid State Ionics*. B. V. R. Chowdari et al. (Eds.). p. 217.

Sharma, J. P. and Sekhon, S. S. (2007). *Solid State Ionics* **178**, 439.

Singh, K. P. and Gupta, P. N. (1998). *Eur. Pol. J.* **34**(7), 1.

Singh, Th. J. and Bhat, S. V. (2004). *Journal of Power Sources* **129**, 280.

Song, J. Y., Wang, Y. Y., and Wan, C. C. (1999). *Journal of Power Sources* **77**, 183.

Souquet, J. L., Levy, M., and Duclot, M. (1994). *Solid State Ionics* **70/71**, 337.

Wagner, J. B. and Wagner, C. (1957). *J. Chem. Phys.* **26**, 1597.

8

Denn, M. M. (1990). Issues in viscoelastic fluid mechanics. *Annu. Rev. Fluid Mech.* **22**, 13–34.

Denn, M. M. (2001). Extrusion instabilities and wall slip. *Annu. Rev. Fluid Mech.* **33**, 265–287.

Evdokia, A., Georgios C. G., and Savvas G. H. (2002). On numerical Simulations of Polymer Extrusion Instabilities. *Appl. Rheol.* **12**, 88104.

Ghaemy, M. and Roohina, S. (2003). Grafting of Maleic Anhydride on Polyethylene in a Homogenous Medium in the Presence of Radical Intiators. *Iranian Polymer Journal* **12**, 2129.

Humberto Palza, Ingo F. C. Naue, and Manfred Wilhelm (2009) In situ Pressure Fluctuations of Polymer Melt Flow Instabilities: Experimental Evidence about their Origin and Dynamics. *Macromol. Rapid Commun.* **30**, 17991804.

Humberto Palza, Susana Filipe, and Ingo F. C. (2010). Naue and Manfred Wilhelm. *Polymer* **51**, 522–534.

Jaewhan Kim, Dong Hak Kim, and Younggon Son, (2009). Rheological properties of long chain branched polyethylene melts at high shear rate. *Polymer* **50**, 49985001.

Lau, H. C and Schowalter, W. R. (1986) A model for adhesive failure of viscoelastic fluids during flow. *Rheol* **30,193206**.

Moayad N. Khalaf, Ali H. Al-Mowali, and Georgius A. Adam. (2008) A Rheological Studies of Modified MAPE Medium Density Polyethylene Blends. *Malaysian Polymer Journal* **3**, 5464.

Ramamurthy, A. V (1986). Wall slip in viscous fluids and influence of materials of construction. *Rheol* **30, 337357**.

Rudolf, J. (1998). Koopmans and Jaap Molenaar. The sharkskin effect in polymer extrusion. *Polymer Engineering and Science* **38**, 101107.

9

Ahmad, M. (2007). *Opt. Commun.* **271**, 457-461.

Barnes, N. P. (1995). Transition Metal Solid-lasers. In *Tunable Lasers Handbook*. F. J. Duarte (Ed.). Academic, New York, pp. 219-291.

Bizet, S., Galy, J., and Gerard, J. F. (2006a). *Macromolecules* **39**, 2574-2583.

Bizet, S., Galy, J., and Gerard, J. F. (2006b). *Polymer* **47**, 8219-8227.

Burroughes, J. H., Bradley, D. D. C., Brown, A. B., Marks, R. N., Mackay, K., Friend, R. H., Burns, P. L., and Holmes, A. B. (1990). *Nature* **347**, 539-541.

Cerdán, L., Costela, A., García-Moreno, I., García, O., Sastre, R., Calle, M., Muñoz, D., and De Abajo, J. (2009). *Macromol. Chem. Phys.* **210**, 1624-1631.

Cerdán, L., Costela, A., García-Moreno, I., Martín, V., and Pérez-Ojeda, M. E. (2011). *IEEE J. Quantum Electron*, (in press).

Cordes, D. B., Lickiss, P. D., and Rataboul, F. (2010). *Chem. Rev.* **110**, 2081-2173.

Costela, A., García-Moreno, I., and Sastre, R. (2001). Materials for solid-state dye lasers. *In Handbook of Advanced Electronic and Photonic Materials and Devices*. H. S. Nalwa (Ed.). Academic, San Diego, CA, vol. 7, pp. 161-208.

Costela, A., García-Moreno, I., and Sastre, R. (2003). *Phys. Chem. Chem. Phys.* **5**, 4745-4763.

Costela, A., García-Moreno, I., Cerdán, L., Martín, V., García, O., and Sastre, R. (2009). *Adv. Mat.* **21**, 4163-4166.

Costela, A., García-Moreno, I., Del Agua, D., García, O., and Sastre, R. (2007). *J. Appl. Phys.* **101**, 073110.

Costela, A., García-Moreno, I., Figuera, J. M., Amat-Guerri, F., Mallavia, R., Santa-María, R. D., and Sastre, R. (1996). *J. Appl. Phys.* **80**, 3167-3173.

Costela, A., García-Moreno, I., Gómez, C., Sastre, R., Amat-Guerri, F., Liras, M., López Arbeloa, F., Bañuelos, J., and López Arbeloa, I. (2002). *J. Phys. Chem.* **A106**, 7736-7742.

Duarte, F. J. (1994). *Appl. Opt* **33**, 3857-3860.

Duarte, F. J. and James, R. O. (2003). *Opt. Lett.* **28**, 2088-2090.

Duarte, F. J. and James, R. O. (2004). *Appl. Opt.* **43**, 4088-4090.

Garnier, F., Hajlaoui, R., Yassar, A., and Srivastava, P. (1994). *Science* **265**, 1684-1686.

García, O., Garrido, L., Sastre, R., Costela, A., and García-Moreno, I. (2008). *Adv. Funct. Mater.* **18**, 2017-2025.

García, O., Sastre, R., García-Moreno, I., Martín, V., and Costela, A. (2008). *J. Phys. Chem. C* **112**, 14710-14713.

Gómez-Romero, P. and Sánchez, C. (2004). *Functional Hybrid Materials*. P. Gómez-Romero and C. Sánchez (Eds.). Wiley-VCH, Weinheim Germany.

Huynh, W. U., Dittmer, J. J., and Alivisatos, A. P. (2002). *Science* **295**, 2425-2427.

In Ref. [40], we studied the photophysics of PM567 and other dyes in six different solvents and only published as an example the curves corresponding to solutions of PM567 in methanol. For the calculations here reported the unpublished but similar curves of PM567 in ethyl acetate have been utilized.

Kanicky, J. (1986). In *Handbook of Conducting Polymers*. T. A. Skotheim (Ed.). Marcel Dekker, New York, p. 543.

Kopesky, E. T., Haddad, T. S., Cohen, R. E., and McKinley, G. H. (2004). *Macromolecules* **37**, 8992-9004.

Lickiss, P. D. and Rataboul, F. (2008). *Adv. Organomet. Chem.* **57**, 1-116.

Markovic, E., Clarke, S., Matisons, J., and Simon, G. P. (2008). *Macromolecules* **41**, 1685-1692.

Nikogosian, D. N. (1997). *Properties of Optical and Laser Related Materials*. A Handbook. Wiley, New York.

Noginov, M. A. (2005). *Solid- State Random Lasers*. Springer, New York.

Pielichowski, K., Njuguna, J., Janowski, B., and Pielichowski, J. (2006). *Adv. Polym. Sci.* **201**, 225-296.

Polson, R. C., Levina, G., and Vardeny, Z. V. (2000). *Appl. Phys. Let.* **76**, 3858-3860.

Pu, K. Y., Zhang, B., Ma, Z., Wang, P., Qi, X. Y., Chen, R. F., Wang, L. H., Fang, Q. L., and Huang, W. (2006). *Polymer* **47**, 1970-1978.

Rahn, M. D. and King, T. A. (1998). *J. Mod. Opt.* **45**, 1259-1267.

Reisfeld, R. (2004). Sol-Gel Processed Lasers. In *Handbook of Sol-Gel Technology*. S. Sakka (Ed.). Springer, Berlin, vol. 3, pp. 239-261.

Sastre, R. and Costela, A. (1995). *Adv. Mater.* **7**, 198-202.

Sastre, R., Martín, V., Garrido, L., Chiara, J. L., Trastoy, B., García, O., Costela, A., and García-Moreno, I. (2009). *Adv. Funct. Mat.* **19**, 3307-3316.

Takeda, S. and Obara, M. (2009). *Appl. Phys. B* **94**, 443-450.

Van de Hulst, H. C. (1981). *Light Scattering by Small Particles*. Dover, New York.

Wiersma, D. S. and Lagendijk, A. (1996). *Phys. Rev.* **E54**, 4256-4265.

Wu, X. H., Yamilov, A., Noh, H., Cao, H., Seeling, E. W., and Chang, R. P. H. (2004). *J. Opt. Soc. Am.* **B21**, 159-167.

Zhao, C. B., Yang, X. J., Wu, X. H., Liu, X. H., Wang, X., and Lu, L. D. (2008) *Polym. Bull.* **60**, 495-505.

Zheng, L., Waddon, A. J., Farris, R. J., and Coughlin, E. B. (2002). *Macromolecules* **35**, 2375-2379.

10

Alivisatos, A. P., Harris, T. D., Carroll, P. J., Stiegerwald, M. L., and Brus, L. E. (1989). Electron–vibration coupling in semiconductor clusters studied by resonance Raman spectroscopy. *J. Chem. Phys.* **90**, 3463.

Arterton, B., Brightwell, J. W., Mason, S., Ray, B., and Viney, I. V. F. (1992). Impact of phase concentrations on structure and electroluminescence of ZnS:Cu. *J. Cryst.Growth.* **117**, 1008.

Becker, W. G. and Bard, A. J. (1983). Photoluminescence and photoinduced oxygen adsorption of colloidal zinc sulfide dispersions. *J. Phys. Chem.* **87**, 4888-4893.

Bellotti, E., Brennen, K. F., Wang, R., and Ruden, P. P. (1988). Monte Carlo study of electron initiated impact ionization in bulk zincblende and wurtzite phase ZnS. *J. Appl. Phys.* **83**(9), 4765.

Bhatti, H. S., Sharma, R., Verma, N. K., Kumar, N., Vadera, S. R., and Manzoor, K. (2006). Lifetime shorting in doped ZnS nanophosphors. *J. Phys. D* **39**, 1754

Biswas, S., Kar, S., and Choudhary, K. S. (2006). Synthesis and Characterization of Zinc Sulfide Nanostructures. *J. Synthesis of Reactivity in Inorganic mMetal oOrganic and Nanometal Chem.* **36**, 33.

Bol, A. A. and Meijerink, A. (1998). Long-lived Mn^{2+} emission in nanocrystalline $ZnS:Mn^{2+}$. *Phys. Rev. B* **58**, 15997-16000.

Bruchez, M., Moronne, M., Gin, P., Weiss, S., and Alivisatos. A. P. (1998). Semiconductor nanocrystals as fluorescent biological labels. *Science* **281**, 2013-2016.

Brus, L. E. (1983). A simple model for the ionization potential of small semiconductor crystallites. *J. Chem. Phys.* **79**(1), 5566-5571.

Brus, L. E. (1984). Electron-electron and electron-hole interaction in small semiconductor crystallites: The size dependence of the lowest excited electronic state. *J. Chem. Phys.* **80**(9), 4403-4409.

Chen, L., Zhang, J., Luo, Y., Lu, S., and Wang, X. (2004). Effect of Zn^{2+} and Mn^{2+} introduction on the luminescent properties of colloidal $ZnS:Mn^{2+}$ nanoparticles. *Appl. Phys. Lett.* **84**, 112.

Colvin, V. L., Schlamp, M. C., and Alivisatos, A. P. (1994). Light-emitting diodes made from cadmium selenide nanocrystals and a semiconducting polymer. *Nature* **370**, 354-357.

Comor, M. I. and Nedeljkovic, J. M (1999). Enhanced photocorrosion stability of colloidal cadmium sulfide-silica nanocomposites. *J. Mater. Sci. lLetter.* **18**(19), 1583-1585.

Davis, M. J. and Williams, R.H. (1989). *In Electroluminescence*. S. Shiohoya. and H. Kobayashi (Eds.). Springer, Berlin, p. 301.

Dhas, N. A., Zaban, A., and Gedanken, A. (1999). Surface synthesis of zinc sulfide Nanoparticle on silica micro-spheres sonochemical prepration, characterization and optical properties. *Chem. Mater.* **11**, 806.

Efros, Al. L. and Efros, A. L., (1982). Interband absorption of light in semiconductor sphere. *Sov. Phys. Semicond.* **16**, 772-776.

Everett, D. H. (1988). *Basic Principles of Colloidal Science*. Royal Society of Chemistry, London, pp. 26-27.

Ghosh, G., Naskar, M. K., Patra, A., and Chatterjee, M. (2006). Synthesis and characterization of PVP-encapsulated ZnS nanoparticles. *J. Opt. Mat.* **28**, 1047-1053.

Guinier, A. (1963). *X-ray Diffraction in Crystals, imperfect crystals and amorphous bodies*. Freeman, San Francisco, CA USA.

Hebalkar, N., Lobo, A., Sainkar, S. R., Pradhan, S. D., Vogel, W., Urban, J., and Kulkarni, S. K. (2001). Properties of zinc sulphide nanoparticles stabilized in silica. *J. Mater. Sc.* **36**, 4377.

Herron, N., Wang, Y., and Eckert, H. (1990). Synthesis and Characterization of Surface-Capped Size-Quantized CdS Cluser. Chemical Control Cluster Size. *J. Am. Chem. Soc.* **112**, 1322-1326.

Hong, P. D., Chen, J. H., and Wu, H. L. (1998). Solvent Effect on Structural Change of Poly (vinyl alcohol). *J. Appl. Polym. Sci.* **69**, 2477-2486.

Huynh, W. U., Dittmer, J. J., and Alivisatos, A. P. (2002). Hybrid nanorod-polymer solar cells. *Science* **295**(5564), 2425-2427.

Jaehun, C., Sunbae, L., and Du-Jeon, J. (2002). ZnS nanoparticle treatment to enhance its luminescence, shape, and stability. *Coree, Repub― liquede, Korea- Japan Forum, Seoul* **377**, 85-88.

Kale, S., Gosavi, S. W., Urban, J., and Kulkarni, S. K., (2006). Nanoshell particles: Synthesis properties and applications. *Current Science* **91**(8), 1038-1052.

Karar, N., Singh, F., and Mehta, B. R. (2004) Structure and photoluminescence of ZnS:Mn nanoparticles. *J. Appl. Phys.* **95**(2), 656.

Kayanuma, Y. (1988). Quantum size effects of interfacing electrons and holes in semiconductor micricrystals with spherical shape. *Phys. Rev. B* **38**(14), 9797-9805.

Kubo, T., Isobe, T., and Senna, M. (2002). Enhancement of Photoluminescence of ZnS:Mnnanocrystals modified by surfactants with phosphate or carboxyl groups via a reverse micelle method. *J. Lumin.* **99**, 39-45.

Kumbhojkar, N., Nikesh, V. V., and Kshirsagar, A. (2000). Photphysical properties of ZnS nanocrystals. *J. Appl. Phys.* **88**(11), 6260-6264.

Malik, M. A, O'brien, P., and Revaprasadu, N. (2001). Synthesis of TOPO-capped Mn-doped ZnS and CdS quantum dots. *J. Mater. Chem.* **11**, 2382-2386.

Manzoor, K., Vadera S. R., Kumar, N., and Kutty T. R. N. (2003). Synthesis and photoluminescence properties of ZnS nanocrystals doped with copper and halogen. *Mater. Chem. Phys.* **82**, 718-725.

Manzoor, K., Vadera, S. R. Kumar, N., and Kutty, T. R. N. (2004). Multicolor electroluminescent devices using doped ZnS nanocrystals. *Appl. Phys. Lett.* **84**, 284.

Martinez-Caston, G. A., Sanchez-Lirede, M. G., Martinez-Mandoza, J. R., Oritega-Zarzosa, G., and Facundo, R. (2005). Characterization of silver sulfide nanoparticles synthesized by a simple precipitation method. *Matter. Lett.* **59**(4), 529-534.

Matijevic, E. (1986). Monodispersed colloids: Art and Science. *Langmuir* **2**, 12.

Nanda, K. K. Sarangi, S. N., and Sahu, S. N. (1998). CdS nanocrystalline films: composition, surface, crystalline size, structural and optical absorption studies. *Nanostr. Mater.* **10**, 1401-1410.

Nanda, .K. K., Sarangi, S. N., and Sahu, S. N. (1999). Visible light emission from CdS nanocrystals. *J. Phys.: D., Appl. Phys.* **32**, 2306-2310.

Pschenitzka, F. and Sturm, J. C. (2001). Excitation mechanisms in dye doped organic light-emitting devices. *Appl. Phy. Lett.* **79**(26), 4354.

Rathore, K., Patidar, D., Janu, Y., Saxena, N. S., Sharma, K., and Sharma, T. P. (2008). Structural and optical characterization of chemical synthesized ZnS nanoparticles. *Chelcogenides letter* **5**, 105.

Rincon, M. E., Martinez, M. W., and Mirand-Hernandez, M. (2003) Structural, optical and photoelectrochemical properties of screen-printed and sintered $(CdS)_x(ZnS)_{1-x}$ $(0 < x < 1)$ films. *Sol. Energy Mater. Sol. Cells* **77**, 25-40.

Rosseti, R., Hull, R., Gibson, J. M., and Brus, L. E. (1985). Excited electronic states and optical spectra of ZnS and CdS crystallite in ~ 15 to 50 A size range: Evolution from molecular to bulk semiconducting properties. *J. Chem. Phys.* **82**(1), 552-59.

Roth, W. L. (1967). *In Physics and Chemistry of II-VI compounds*. M. V. Aven and J. S. Perner (Eds.). North-Hollend Publishing Co., Amsterdem, p. 124.

Sharma, R. and Bhatti, H. S. (2007). Photoluminescence decay kinetics of doped ZnS nanophosphors. *Nanotechnology*, **18**, 465703.

Soo, Y. L., Ming, Z. H., Huang, S. W., Kao, Y.H., Bhargava, R. N., and Gallagher, D. (1994). Local structure around in Mn luminescent centers in Mn doped nanocrystals of ZnS. *Phys. Rev. B* **50**, 7602

Sooklal, K., Cullum, B. S., Angel, S. M., and Murphy, C. J. (1996). Photophysical properties of ZnS nanoclusters with spatially localized Mn^{2+}. *J. Phy. Chem.* **100**(11), 4551–4555.

Spanhel, I. and Anderson, M. A. (1991). Semiconductor clusters in the sol-gel process: quantized aggregation, gelation, and crystal growth in concentrated zinc oxide colloids. *J. Am. Chem. Soc.* **113**, 2826-2833

Tessler, N., Medvedev, V., Kazes, M., Kan, S., and Banin, U. (2002). Efficient near-infrared polymer nanocrystal light-emitting diodes. *Science* **295**, 1506-1508.

Wageh, S., Liu, S. M., Fang, T. Y., and Xu, X. R. (2003). Optical properties of strong luminescing mercaptoacetic acid capped ZnS nanoparticle. *J. Lumin.* **102**,768-773.

Wang, Y. and Herron, N. (1992). Photoconductivity of CdS nanocluster-doped polymers. *Chem. Phys. Lett., 1992*, **200**, 71-75.

Yakimov, A. I., Stepina, N. P., Dvurechenskii, A. V., and Scherbakova, L. A. (1995). Low dimensional hopping conduction. *Physica B*, **205**, 298-304.

Yang, H., Wang, Z., Song, L., Zao, M., Chen, Y., Dou, K., Yu, J., and Wang, L. (1997). Study of optical properties of manganese doped ZnS nanocrystals. *Mater. Chem. Phys.* **47**, 249-251.

Yang, Y., Huang, J., Liu, S., and Shen, J. (1997). Preparation, characterization and electroluminescence of ZnS nanocrystals in polymer matrix. *J. Mater. Chem.* **7**(1), 131-133.

Index

D

E

G

H

K

L

R

S

T

V

Z